FORMALDEHYDE

FORMALDEHYDE
Toxicology • Epidemiology • Mechanisms

John J. Clary

Celanese Corporation
New York, New York

James E. Gibson

Chemical Industry Institute of Toxicology
Research Triangle Park, North Carolina

Richard S. Waritz

Hercules Incorporated
Wilmington, Delaware

MARCEL DEKKER, INC. New York and Basel

Library of Congress Cataloging in Publication Data

Main entry under title:

Formaldehyde, toxicology, epidemiology, and mechanisms.

Papers of a meeting held Nov. 3, 1982, sponsored by the Formaldehyde Institute.
Includes index.
1. Formaldehyde--Toxicology--Congresses. I. Clary, John J. II. Gibson, James E. III. Waritz, Richard S. IV. Formaldehyde Institute.
RA1242.F6F674 1983 615.9'5136 83-14273
ISBN 0-8247-7025-0

MARCEL DEKKER, INC.
270 Madison Avenue, New York, New York 10016

Current printing (last digit):
10 9 8 7 6 5 4 3 2 1

PRINTED IN THE UNITED STATES OF AMERICA

PREFACE

Formaldehyde is a chemical of great metabolic, medical, societal and industrial importance. It is a truly ubiquitous chemical and a thorough understanding of any health hazards from it is extremely important to society. It has long been known that high concentrations of formaldehyde are irritating to man and that it can cause skin sensitization. The Chemical Industry Institute of Toxicology (CIIT) reported a study in December, 1981 which showed that chronic exposure to cell-damaging concentrations of formaldehyde caused a nasal cancer in rodents. Full elucidation of the toxic effects, mechanisms of toxicity and hazard from exposure to exogenous sources of this metabolite will almost certainly require advancing the state of the art in toxicology. Sound regulation of exposure to exogenous formaldehyde will require understanding of biochemical and toxicological principles as well as the results of individual experiments.

Because of formaldehyde's importance, the CIIT sponsored a conference in November, 1980 in which current research on formaldehyde was presented. The meeting brought scientists and regulators together to hear and discuss this spectrum of then-current research. A second meeting, sponsored by the Formaldehyde Institute, was held on November 3, 1982 to provide a forum for scientists to report toxicological study results obtained since the 1980 conference. This book contains the papers presented at the November 3, 1982 scientific meeting.

J. J. Clary, Ph. D.
J. E. Gibson, Ph. D.
R. S. Waritz, Ph. D.

CONTENTS

Preface iii

Introduction vii

Contributors xiii

Acknowledgments xvi

1. Occupational Exposure to Formaldehyde - Recent NIOSH Involvement 1
Leo M. Blade

2. Mathematical Cancer Risk Assessment for Formaldehyde 31
Frank W. Carlborg

3. Case Control Study of Cancer Deaths in Du Pont Workers with Potential Exposure to Formaldehyde 47
William E. Fayerweather, Sidney Pell, and Joel R. Bender

4. Mortality of Ontario Undertakers: A First Report 127
Richard J. Levine, Dragana A. Andjelkovich, Linda K. Shaw, and R. Daniel DalCorso

5. Skin Initiation/Promotion Study with Formaldehyde in Sencar Mice 147
E. Frederick Spangler and Jerry M. Ward

6. Skin Initiation/Promotion Study with Formaldehyde in CD-1 Mice 159
Neil D. Krivanek, Nancy C. Chromey, and John W. McAlack

7. Mutagenic Effects of Formaldehyde in Bacterial and Human Cells 173
Victor S. Goldwater, Punya Temcharoen, and William G. Thilly

8. Formaldehyde and the Nasal Mucociliary Apparatus 193
Kevin T. Morgan, Daniel L. Patterson, and Elizabeth A. Gross

9. Reaction of Formaldehyde in the Rat Nasal Mucosa 211
Henry d'A. Heck and Mercedes Casanova-Schmitz

10. The Effect of Formaldehyde Exposure in Cytotoxicity and Cell Proliferation 225
James A. Swenberg, Elizabeth A. Gross, Holly W. Randall, and Craig S. Barrow

11. Mechanisms of Formaldehyde Toxicity and Risk Evaluation 237
Thomas B. Starr

Conference Summary and Perspectives 259

Index 277

INTRODUCTION

The dose-response curve and the no-observable-effect level greater than zero have been givens in toxicology since before the time of Paracelsus. These concepts have served society well for classical toxicological effects from food additives and industrial chemicals.

Historically, the problem of extrapolation to humans from the no-observable-effect level in animals has been handled in two different ways in these two areas. Because there is no control over dietary habits, larger safety factors have been applied to extrapolate animal test data for food additives than have been used for industrial chemicals. The rationale is that exposure of the worker to industrial chemicals can be better controlled than dietary habits. For an occasional maverick chemical, the toxicological data base was inadequate and illness due to the chemical appeared in the exposed population. The illness generally appeared more or less quickly and usually could be traced to the cause. As a result, the number of people injured was usually minimal, as remedial steps could be taken quickly. Fortunately, most classical toxic effects that manifest themselves under these conditions spontaneously repaired shortly after exposure reduction, or could be kept in remission by exposure reduction, e.g., sensitization. Also fortunately, the toxic effect caused by these chemicals at these doses was not death.

Carcinogens may or may not heed both these givens of dose response and no-observable-effect level. In addition, cancer has a long latent period and it may not manifest itself for 20 or more years after exposure initiation. Many people could be exposed before a chemical is found to be carcinogenic in humans. In addition, this toxic effect usually can not be reversed by reduction or elimination of exposure, and the usual end result is death. Thus, inappropriate toxicological judgments in this area of toxicology can be extremely costly to individuals and society. As with classical toxicological effects, tumorigens/carcinogens have been handled differently, at least in the U.S., depending on whether they were a food additive or an industrial chemical. Chemicals found to be carcinogenic in animals were simply prohibited in U.S. food. Thus, for U.S. foods, questions regarding the shape of the lower end of the dose-response curve and the presence or absence of a no-observable-effect level as well as the need for extrapolating animal test results to man do not exist.

Many toxicological facts do not support this no-contact philosophy. These facts have been recognized and utilized by toxicologists and other health professionals for setting human exposure levels greater than zero for animal carcinogens. Exposures at these levels for decades have not resulted in any observable adverse health effects. The American Conference of Industrial Hygiene (ACGIH) has assigned Threshold Limit Values (TLV's) even to acknowledged human carcinogens.

Some of the facts supporting an no-observable-effect level for human carcinogens are: 1) Even though sunlight is carcinogenic, not everyone exposed to sunlight gets skin cancer, 2) there does not seem to be an incidence of liver tumors in industrialized countries attributable to the unavoidable presence of small amounts of aflatoxin in food stuffs even though higher levels in other countries are associated with a liver tumor increase 3) some chlorinated industrial solvents are hepatocarcinogenic at high doses in the $B_6C_3F_1$ mouse, but have given no indication of carcinogenicity in humans, even though there has been low or moderate exposure of humans for 20 to 30 years, 4) some chemicals seem to be carcinogenic in animals only in the presence of a chronic irritation; for example, the bladder cancer subsequent to chronic bladder irritation by stones and concurrent chronic administration of nitrilotriacetic acid, and 5) even though saccharin was carcinogenic in animals at high doses there does not seem to be any convincing indication that it is carcinogenic in man at human use levels.

One must also consider the probable predictive value of any animal test; that is, the predictive value of any positive response as indicating a potential or possible positive response in humans. Few would accept the results of a positive implant study of a polymeric or glass sheet as indicative of human risk at home or at work. Similarly, most toxicologists discount tumors at the injection site in a subcutaneous carcinogenicity study. As mentioned above, other test results also do not seem to be very predictive of risk to man.

In the case of formaldehyde, cancer at the site of irritation has been found in some rodent species only following chronic inhalation at levels that would be very irritating to humans. This very irritation from formaldehyde vapor at low atmospheric concentrations is a self-limiting factor which tends to restrict human exposure to low, non-irritating, concentrations.

In living things, including man, formaldehyde also is a normal metabolite and is the component of a major essential metabolic system -- the one-carbon pool. In the case of this animal carcinogen, there is some normal metabolic mechanism for handling it in all plant and animal species. This suggests the reality of a no-effect level. The data reported by CIIT also suggest a dose-response relationship for carcinogenicity in rats following chronic inhalation of formaldehyde.

Assuming that formaldehyde is carcinogenic to man at some high level, there is still the problem of extrapolating the animal test results to man. What is the human dose equivalent to

15 or 6 ppm in the rat? What is the shape of the dose-response curve for formaldehyde in humans? What is the shape of the curve in rats below 2 ppm? Are the rat and mouse data from the CIIT study really inconsistent as a predictor for man?

Formaldehyde is a major chemical building block in our society. Its outright ban would cause dramatic changes in our society. Adoption of a requirement for zero contact would also have financial consequences that would result in severe negative consequences to our society. But that societal impact cannot be used to justify exposure of members of our society to a hazardous concentration of formaldehyde.

Society also cannot afford a schizoid approach -- an approach that ignores animal carcinogenicity studies for some chemicals, such as saccharin, that have a popular appeal and for some basic chemicals that are fundamental chemical building blocks, such as formaldehyde, and prohibits the use of all other chemicals that are active in animal studies. A responsible, consistent approach is necessary; one that is responsible to society and also responsible to individuals contacting the chemical.

The cost of reducing exposure varies inversely with the atmospheric level desired -- at least for chemicals with a reasonable volatility. Zero exposure can be achieved, such as it has been for the biologically dangerous radioactive isotopes, but at an enormous cost in dollars and productivity. If zero contact is necessary for health reasons, then it should be sought. But if it is not necessary, then it would seem to be a waste of societal resources to try to achieve it.

A response consistent with real human risk requires that scientists continue to develop information on mechanisms of carcinogenicity, because the history of nature is such that there probably will be more than one mechanism. We must continue to investigate each chemical that is positive in an animal carcinogen study to determine if that result is metabolically relevant to humans. We must continue to investigate the carcinogenicity assays themselves, so that societal efforts are not wasted in control based on tests that may be irrelevant.

Hopefully, investigations into the unique aspects of the mechanisms of carcinogenicity of each animal carcinogen will provide the information necessary to translate that information into risk for man and an exposure level for that specific chemical, physical agent or combination of exposures, which will not increase the real risk of carcinogenicity for man.

This conference was held so that scientists working to understand the toxicologic effects of formaldehyde in animals and to translate those effects into a judgment on risk to man could learn of the work of other scientists and talk freely with these other scientists. It also was held in the hope that this exchange would speed the acquisition of knowledge that would lead to control of human exposure to formaldehyde that was responsible both to our society's living standards and to the individual members of society contacting it.

CONTRIBUTORS

DRAGANA A. ANDJELKOVICH, Department of Epidemiology, Chemical Industry Institute of Toxicology, Research Triangle Park, North Carolina

CRAIG S. BARROW, Department of Pathology, Chemical Industry Institute of Toxicology, Research Triangle Park, North Carolina

JOEL R. BENDER, E. I. du Pont de Nemours & Co., Inc., Chattanooga, Tennessee

LEO M. BLADE, National Institute of Occupational Safety and Health, Robert A. Taft Laboratories, Cincinnati, Ohio

FRANK W. CARLBORG, St. Charles, Illinois

MERCEDES CASANOVA-SCHMITZ, Departments of Pathology, General, and Biochemical Toxicology, Chemical Industry Institute of Toxicology, Research Triangle Park, North Carolina

NANCY C. CHROMEY, Central Research and Development Department, Haskell Laboratory, E. I. du Pont de Nemours & Co., Inc., Newark, Delaware

R. DANIEL DALCORSO, Department of Epidemiology, Chemical Industry Institute of Toxicology, Research Triangle Park, North Carolina

WILLIAM E. FAYERWEATHER, Medical Department, E. I. du Pont de Nemours & Co., Inc., Wilmington, Delaware

VICTOR GOLDMACHER, Department of Nutrition and Food Science, Toxicology Group, Massachusetts Institute of Technology, Cambridge, Massachusetts

ELIZABETH A. GROSS, Department of Pathology, Chemical Industry Institute of Toxicology, Research Triangle Park, North Carolina

HENRY d'A. HECK, Departments of Pathology, General, and Biochemical Toxicology, Chemical Industry Institute of Toxicology, Research Triangle park, North Carolina

NEIL D. KRIVANEK, Central Research and Development Department, Haskell Laboratory, E. I. du Pont de Nemours & Co., Inc., Newark, Delaware

RICHARD J. LEVINE, Department of Epidemiology, Chemical Industry Institute of Toxicology, Research Triangle Park, North Carolina

JOHN W. McALACK, Central Research and Development Department, Haskell Laboratory, E. I. du Pont de Nemours & Co., Inc., Newark, Delaware

KEVIN T. MORGAN, Department of Pathology, Chemical Industry Institute of Toxicology, Research Triangle Park, North Carolina

DANIEL L. PATTERSON, Department of Pathology, Chemical Industry Institute of Toxicology, Research Triangle Park, North Carolina

SIDNEY PELL, Employee Relations Department, Medical Division, E. I. du Pont de Nemours & Co., Inc., Wilmington, Delaware

HOLLY W. RANDALL, Department of Pathology, Chemical Industry Institute of Toxicology, Research Triangle Park, North Carolina

LINDA K. SHAW, Department of Epidemiology, Chemical Industry Institute of Toxicology, Research Triangle Park, North Carolina

E. FREDERICK SPANGLER, Department of Chemical Carcinogenesis, Microbiological Associates, Inc., Bethesda, Maryland

THOMAS B. STARR, Department of Epidemiology, Chemical Industry Institute of Toxicology, Research Triangle Park, North Carolina

JAMES A. SWENBERG, Department of Pathology, Chemical Industry Institute of Toxicology, Research Triangle Park, North Carolina

PUNYA TEMCHAROEN, Department of Nutrition and Food Science, Toxicology Group, Massachusetts Institute of Technology, Cambridge, Massachusetts

WILLIAM G. THILLY, Department of Nutrition and Food Science, Toxicology Group, Massachusetts Institute of Technology, Cambridge, Massachusetts

JERROLD M. WARD, Laboratory of Comparative Carcinogenesis, National Cancer Institute, Frederick Cancer Research Facility, Frederick, Maryland

ACKNOWLEDGMENTS

The editors would like to thank all of the scientists who participated in this conference. We also thank the Formaldehyde Institute for sponsoring this conference, for providing the funding, and for handling the business management of the conference; Mrs. Barbara Burr of the Word Processing Center at Hercules Incorporated for retyping all of the manuscripts and putting them in a consistent format; and Lori Gialleonardo, Rosalind Magilton, and Bill Smock of the Formaldehyde Institute staff for their assistance in transforming the plans for the conference into reality.

FORMALDEHYDE

CHAPTER 1

OCCUPATIONAL EXPOSURE TO FORMALDEHYDE-RECENT NIOSH INVOLVEMENT

L. M. Blade

National Institute for Occupational Safety and Health
Cincinnati, Ohio

During the past two years, the NIOSH has been at the forefront of industrial hygiene activity relating to occupational exposures to formaldehyde. It has conducted several industrial hygiene surveys in which air samples for formaldehyde were collected. Some of the higher concentrations of formaldehyde recorded were found in a paper manufacturing facility, a biological laboratory, and a facility which packages paraformaldehyde in packets for retail sale as an anti-mildew agent. The institute has also conducted industrial hygiene monitoring in textile printing and finishing, garment manufacturing, and dialysis facilities. Summaries of the findings from these and other surveys are presented. NIOSH's Division of Physical Sciences and Engineering has developed a new solid-sorbent-tube sampling and analytical procedure, Physical and Chemical Analytical Method No. 354, in which formaldehyde is sampled with Chromosorb 102[R] coated with N-benzylethanolamine. A detailed description of this recent advance, published in August 1981, is presented.

INTRODUCTION

The National Institute for Occupational Safety and Health (NIOSH) is involved extensively with the important topic of occupational exposure to formaldehyde, and it is the purpose of this paper to bring the reader up to date on this involvement by relating some of the important studies which NIOSH has conducted on formaldehyde exposure since the Chemical Industry Institute of Toxicology's (CIIT's) 1980 conference on formaldehyde.

During these past two years, NIOSH has conducted many industrial hygiene surveys in which formaldehyde exposure levels were measured; the results of many of these surveys will be summarized here. In addition, NIOSH has developed a new sampling and analytical method for the quantification of formaldehyde in air; brief descriptions of this new method, and of four other methods used in the industrial hygiene surveys, are included. The sampling and analytical methods are described first, and a discussion of the results of the industrial hygiene surveys follows. It is important to note that the descriptions of the sampling methods contain references to brand-name products and companies, but no endorsement by NIOSH of the products is intended.

METHODS

Sampling and Analytical

Five methods of sampling the workplace air for formaldehyde were used in the industrial hygiene surveys (Table 1). The first of these is the Draeger® Detector Tube for formaldehyde. This well-known method incorporates a specially designed hand-held pump which draws a short-term air sample through a glass detector tube; a color change in the sorbent provides a direct-reading measurement. This method has a lower detection limit of 0.5 ppm,

TABLE 1

Air Sampling Methods Employed by NIOSH

Method	Area or Pers.	Sample Time
Detector Tube (Draeger[R])	Area	Inst.
CEA Instruments Model 555[R]	Area	Contin.
Sodium-Bisulfite Impinger	Area	1 Hr.
Impregnated-Charcoal Tube	Both	1-8 Hr.
Coated-Chromosorb 102[R] Tube	Both	4 Hr.

and is suitable for area sampling, particularly for range-finding. The second method used incorporates the CEA Instruments Model 555[R] direct-reading ambient-air monitor. This device uses a modified pararosaniline colorimetric method to continuously monitor for formaldehyde. An LED display gives instantaneous concentration readings, and a strip-chart record provides a permanent record of these concentrations. This device is bulky and relatively expensive and, like the Draeger[R] Detector Tube, is suitable only for area sampling. The third method incorporates a portable air pump drawing 1.0 liters of air per minute (Lpm) through a midget impinger containing an aqueous solution of 1% sodium bisulfite. This solution is analyzed for formaldehyde by the chromotropic-acid colorimetric method in accordance with NIOSH-recommended Physical and Chemical

Analytical Method (P&CAM) 125.[1] This method is most useful for area sampling, as opposed to personal sampling, because it incorporates a liquid collection medium and because the recommended sampling time is only 1 hour. NIOSH has evaluated this method for the range of 0.1 to 2.0 ppm, and a precision of ±5% was documented. The fourth sampling method employed for the NIOSH surveys incorporates a glass tube, packed with specially impregnated charcoal, through which air is sampled at a flow rate of 200 cc/min or 1.0 Lpm, depending upon the expected concentration range. For analysis, aqueous 0.1% hydrogen peroxide solution is used to desorb the formaldehyde from the sorbent. The resulting solution is analyzed for formaldehyde using ion chromatography in accordance with NIOSH P&CAM 318.[2] This air sampling method was designed for convenient personal breathing-zone sampling for 8-hour shifts. However, problems with sample stability were identified in that samples in storage were prone to the loss of analyte, apparently due to time and temperature effects.

In 1981, NIOSH developed a new solid-sorbent sampling method, and the impregnated-charcoal-tube method discussed above is no longer recommended. The new NIOSH method uses a solid-sorbent-filled tube and is therefore suitable for long-term personal sampling. The glass sample tube contains Chromosorb 102[R] (or XAD-2 Resin[R]) coated with N-benzylethanolamine. NIOSH uses a Dupont Model P30A[R] personal air-sampling pump, calibrated to an air-flow rate of 50 cc/min, to draw air though

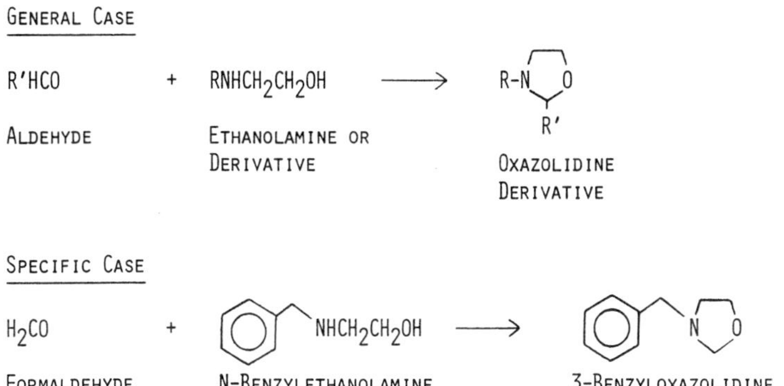

Fig. 1. Chemistry of Coated Chromasorb (XAD-2) tube sampling method.

the sorbent tube. Formaldehyde in the sampled air reacts with the N-benzylethanolamine coating to form 3-benzyloxazolidine. Figure 1 illustrates the chemistry of this reaction. The first reaction shown is the general case for this class of reactions: any aldehyde can react with ethanolamine or its derivatives to form various oxazolidine derivatives. The second reaction is the specific reaction incorporated into this method: N-benzylethanolamine reacts with formaldehyde to form 3-benzyloxazolidine. For analysis, the 3-benzyloxazolidine is desorbed from the sorbent with isooctane. The resulting solution is analyzed for 3-benzyloxazolidine using capillary-column gas chromatography with flame ionization detection, in accordance with NIOSH P&CAM 354.[3].

This new sampling and analytical method is specific for formaldehyde. The oxazolidines formed with acetaldehyde, propionaldehyde, and <u>n</u>-butyraldehyde do not interfere. However, a compound with the same retention time as 3-benzyloxazolidine will present a positive interference. The method has been evaluated over the range of 0.45 to 3.84 parts per million (ppm) for a 12 L sample. At a flow rate of 50 cc/min, a 12 L sample can be collected in 4 hours. The major improvement of this method over the impregnated-charcoal-tube method is the high stability of the sample in storage. Whereas charcoal-tube samples in storage were prone to the loss of analyte, samples collected with Chromosorb or XAD-2 tubes can be stored at least 14 days at room temperature before analysis without loss of the analyte.

Before moving on to a discussion of the industrial hygiene surveys, it is important to make one comment regarding some of the data which will be presented. Many of the samples were collected with the impregnated-charcoal-tube method. Although some problems with this method have been identified, and NIOSH does not currently recommend its use, it does not necessarily follow that the results of all samples collected using this method must be ignored. Often the results correlate fairly well with those of other sampling methods. However, it does mean that results obtained with this method should be viewed with some skepticism if no other methods were utilized simultaneously. In cases where this method gives results far different from those obtained from another method, then the charcoal-tube results probably should not be used.

The charcoal-tube method probably was useful as a range-finding device, and most of the results presented herein which were obtained with this method are valuable from that standpoint.

INDUSTRIAL HYGIENE SURVEYS

When the data collected during the industrial hygiene surveys were compiled and assembled, it became evident that three exposure groups were present. It was found that the data could be grouped based on the levels of formaldehyde, the sources of formaldehyde, and the types of facilities in which formaldehyde was found. The ranges assigned to the exposure group will be summarized first, then the sources of formaldehyde and the types of facilities in each group will be summarized. Finally, an overview of the data will be presented.

Exposure Groups - Ranges of Concentrations

The ranges assigned to the three exposure groups are summarized as follows (Table 2): The levels of formaldehyde found in the facilities of the high-exposure group were almost all above 1.0 ppm, and most were between 2.0 and 2.5 ppm. The levels of formaldehyde detected in intermediate-exposure-group facilities were almost all between 0.1 and 1.0 ppm, and most were between 0.2 and 0.7 ppm. The levels of formaldehyde found in locations which have been placed into the low-exposure group were almost all below 0.1 ppm, and they were often around 0.05 ppm. Unfortunately, many of the results included in this last group were below the lowest detection limit of the instrument used.

Sources of Formaldehyde

The known or suspected sources of formaldehyde at the facilities under discussion are summarized below, and in Table 3. The source of formaldehyde in the facilities included in the

TABLE 2

Exposure Groups-
Ranges, Sources, and Types of Facilities

Exposure Group	Characteristic Exposure Levels (PPM)
High	>1.0, Most Between 2.0 and 2.5
Intermediate	0.1 to 1.0, Most Between 0.2 and 0.7
Low	<0.1, Often Around 0.05

TABLE 3

Exposure Groups-
Ranges, Sources, and Types of Facilities

Exposure Group	Sources of Formaldehyde
High (>1.0 ppm)	Frequent or Continuous Use of Formalin
Intermediate (0.1 to 1.0 ppm)	Intermittent Use of Formalin or Off-Gassing or Pyrolizing from Process Materials
Low (<0.1 ppm)	Off-Gassing from Building or Furnishing Materials, or Background (Pollution) Levels

high-exposure group was formalin solution in frequent or continuous use. At the facilities in the intermediate-exposure group, two basic types of sources were seen. One formaldehyde source was formalin used intermittently; the other source was formaldehyde off-gassing or pyrolyzing from process materials. The sources of formaldehyde in the facilities in the low-exposure group generally were suspected to be building or furnishing materials which off-gassed the vapor.

Types of Facilities

The facilities which were surveyed are involved in a variety of activities (Table 4). The high-exposure group consists of two laboratories, one of which was the gross histology laboratory of a hospital[4], and the other of which was a government laboratory[5]. The intermediate exposure group consists of three facilities in which formalin was used only intermittently and eight facilities in which formaldehyde evolved from other process materials. Two of the facilities intermittently using formalin were hospital dialysis units, in which formalin is used to disinfect the machines[6,7]. The other was a government animal-dissecting laboratory[8]. Four of the eight facilities in which the formaldehyde evolved from materials used in the processes were garment manufacturing facilities using crease-resistant cloth treated with formaldehyde-based resins[9-12]; free formaldehyde in this cloth off-gasses into the workroom air. The fifth of these eight facilities was a specialty-chemical manufacturing plant in which formaldehyde was

TABLE 4

Exposure Groups-
Ranges, Sources, and Types of Facilities

Exposure Group	Types of Facilities
High (>1.0 ppm)	Laboratories Frequently Using Formalin
Intermediate (0.1 to 1.0 ppm)	Dialysis Facilities, Dissecting Laboratory, Garment Manuf. Facilities, Chem. Manuf., Hospital, Glass Manuf., Paraformaldehyde Pkg.
Low (<0.1 ppm)	Offices and Public Buildings

one of many chemicals in the workplace environment[13]. The next of these facilities was engaged in the manufacture of scientific glassware; decals were applied to the glassware by heating in an oven, and an analysis of the decomposition products of the decals indicated that formaldehyde, as well as several other compounds, could be evolved[14]. The seventh of these eight facilities was a hospital; the survey was conducted in an area where an oven was used to heat, and thereby soften, acrylic, polypropylene, and polyethylene prior to the formation of these materials into prosthetics. Once again, analysis of the decomposition products indicated that formaldehyde was one of several compounds that could be present[15]. The final facility of this group was engaged in the packaging of paraformaldehyde in gas-permeable paper bags for consumer use as a mildew retardant[16]. The low-exposure group consists of the following seven facilities:

the offices of four private companies[17-20], the administration building of a college[21], one government office building [22], and an elementary school[23]. Confirmed or suspected sources of formaldehyde at these locations included: off-gassing from wood furniture, carpeting, wallpaper, and other furnishings, as well as insulation; and the induction of polluted outside air.

RESULTS

It is suitable at this point to provide a summary of the industrial hygiene data which permitted the grouping of the surveyed facilities into the exposure groups. For the purpose of providing an overview of the results of a large number of industrial hygiene surveys, it was necessary to combine the results of personal-breathing-zone samples and area samples, samples collected with different methods, and in many cases, samples from different facilities. Therefore, the generation of statistical parameters, such as mean and confidence limits, was deemed to be a questionable validity. Instead, the data summaries focus on the overall ranges of the formaldehyde levels measured, as well as the ranges into which the vast majority of the values fall (see Tables 5-8). If greater detail about the industrial hygiene data is desired, the reader is advised to consult Tables 9-18. The results of each type of sample (area or personal), collected by each method, are summarized by ranges and means. These results are organized by survey location and are placed into tables (Tables 9-18) based on type of facility.

TABLE 5

Formaldehyde Exposures in Facilities Frequently Using Formalin
Summary of Air Sampling Results

Location	Personal or Area	No. Samples	Range (ppm)
Hospital	Both	5	1.9-2.3
Gov. Agency	Both	2	0.8-2.4

TABLE 6

Formaldehyde Exposure in Facilities Intermittently Using Formalin
Summary of Air Sampling Results

Facilities	Personal or Area	No. Samples	Range (ppm)
Dialysis/2	Both	22	0.04-0.90, Most Between 0.2 and 0.7
Dissect Lab/1	Both	27	0.05-1.04, Most Between 0.1 and 0.8

At the two facilities in the high-exposure group (the hospital laboratory and the government laboratory), seven air samples for formaldehyde were collected (Table 5). These were personal as well as area samples, and the results ranged from 0.8 to 2.4 ppm, although six of the seven results ranged from 1.9 to 2.4 ppm. The results from the hospital laboratory ranged from 1.9 to 2.3 ppm[4], while those from the government laboratory were 0.8 to 2.4 ppm[5].

TABLE 7

Formaldehyde Exposures Due to Evolution or Pyrolysis From Process Materials-Summary of Air Sampling Results

Type Facility	No. Facilities	No. Samples	Personal or Area	Range (ppm)
Garment Manuf.	4	181	Both	0.05-0.94, Most Between 0.15 and 0.7
Chem. Manuf.	1	8	Both	0.03-1.6
Glass Manuf.	1	3	Both	0.42-0.64
Hospital	1	2	Area	0.37-0.73
Paraformaldehyde Pkging.	1	18	Both	<0.25-3.4, Most Between 0.3 and 1.1

TABLE 8

Formaldehyde Exposures in Offices and Public Buildings Summary of Area Air Sampling Results

Location	No. Samples	Range (ppm)	Mean (ppm)
Law Office	2	0.05-0.06	0.055
College Admin. Bldg.	37	0.02-0.12	0.065
	10	<0.10	-
	3	<0.04	-
Five Others	NA	None Detected	

TABLE 9

Air Sampling Methods Employed by NIOSH

Method	Abbrev.
- Detector Tube (Draeger[R])	DT
- CEA Instruments Model 555[R]	CEA
- Impinger with Sodium Bisulfite Solution	Imp.
- Impregnated-Charcoal Tube	CT
- Coated-Chromosorb 102[R] Tube	Chrom.

TABLE 10

Formaldehyde Exposures in Laboratories Using Formalin
Air Sampling Results

Location	Type	Method	No. Samples	Range (ppm)	Mean (ppm)
Hospital	Personal	Imp.	2	2.2-2.3	2.25
	Personal	CT	1	1.9	-
	Area	CT	2	2.0-2.0	2.0
Gov. Agency	Personal	CT	1	2.4	-
	Area	CT	1	0.8	-

TABLE 11

Formaldehyde Exposures in Dialysis Unit Disinfection
Air Sampling Results

Facility	Type	Method	No. Samples	Range (ppm)	Mean (ppm)
A	Area	CT	7	ND-0.90	0.47
	Personal	CT	3	0.27-0.63	0.44
Follow-up Visit:					
	Personal	Unk.	7	ND	-
B	Personal	CT	2	0.30-0.47	0.38
	Area	CT	2	0.22-0.26	0.24
	Area	CEA	-	0.04-0.50	0.15*

*Time-weighted average (TWA).

TABLE 12

Formaldehyde Exposures in an Animal-Dissecting Lab
Air Sampling Results

Type	Method	No. Samples	Range (ppm)	Mean (ppm)
Personal	Chrom.	15	<0.38-1.04	-
Area	Chrom.	1	0.41	-
Area	Chrom.	2	<0.43-<0.47	-
Area	Imp.	6	0.05-0.40	0.15
Area	CEA	3*	0.11-0.29**	0.18**

*No. of days at same location.
**Range and mean of daily TWA's.

TABLE 13

Formaldehyde Exposures in Garment Manufacturing
Results of Personal Charcoal-Tube Air Samples

Plant	No. Samples	Range (ppm)	Mean (ppm)
A	10*	0.20-0.63*	0.33*
B	20*	<0.14-<0.21*	-
C	10	0.15-0.30	0.23

*8-hour TWA's

TABLE 14

Formaldehyde Exposures in Garment Manufacturing
Results of Area Charcoal-Tube Air Samples

Plant	No. Samples	Range (ppm)	Mean (ppm)
A	9	0.20-0.42	0.26
B	23	<0.08-0.18	-
C	11	0.11-0.24	0.19

Among the 11 facilities in the intermediate-exposure group, 3 (the 2 hospital dialysis units and the animal-dissecting laboratory) were intermittently using formalin (Table 6). The two hospital dialysis units had airborne concentrations of formaldehyde ranging from 0.04 to 0.90 ppm, and most sample results ranged from 0.2 to 0.7 ppm[6,7]. These figures are based

TABLE 15

Formaldehyde Exposures in Garment Manufacturing
Results of Area Impinger Air Samples

Plant	No. Samples	Range (ppm)	Mean (ppm)
A	9	0.18-0.35	0.25
B	23	0.03-0.30	0.15
C	11	0.22-0.40	0.31

TABLE 16

Formaldehyde Exposures in Garment Manufacturing
Results of CEA Continuous Air Monitoring

Plant	No. Locations Sampled	Range (ppm)	TWA (ppm)
A	13	0.39-1.12*	0.69**
B	20	0.05-0.94	0.31
C	8	0.24-0.71	0.46
D	14	0.24-0.44	0.33

* Range of max. values at locations.
**Mean of max. values at locations.

TABLE 17

Formaldehyde Exposures in Other Industries
Air Sampling Results

Facility	Type	Method	No. Samples	Range (ppm)	Mean (ppm)
Chem. Manuf.	Personal	Imp.	1	0.11	-
	Personal	Imp.	2	0.04-1.6	0.8
	Area	Imp.	5	0.03-0.43	0.17
Glass Manuf.	Personal	CT	1	0.42	-
	Area	CT	2	0.45-0.64	0.54
Hospital	Area	Imp.	2	0.37-0.73	0.55
Paraform-maldehyde Pkging.	Personal	Chrom.	10	<0.25-0.85*	0.56*
	Area	CEA	8**	0.28-3.40	1.17**

* Calculated 8-Hr. TWA's.
**No. of locations; mean of TWA's of locations.

on the results of 22 air samples, both area and personal. The animal-dissecting laboratory was found to have exposures ranging from 0.05 to 1.04 ppm, based on 27 air samples; most results were between 0.1 and 0.8 ppm[8].

The remaining eight facilities (four garment-manufacturing facilities, one specialty-chemical manufacturing plant, one hospital, and one paraformaldehyde-packaging plant) in the intermediate-exposure group were found to have airborne concentrations of formaldehyde ranging from 0.03 to 1.6 ppm, based on results of 212 air samples (Table 7). However, almost

TABLE 18

Formaldehyde Exposures in Office and Public Buildings
Area Air Sampling Results

Location	Method	No. Samples	Range (ppm)	Mean (ppm)
Law Office	Imp.	2	0.05-0.06	0.055
College Admin. Bldg.	Imp.	37	0.02-0.12	0.065
	Imp.	10	<0.10	-
	CT	3	<0.04	-
Telephone Co.	CT	6	ND	ND*
Office Bldg.	DT	NA	<0.5	
Sales Office	DT	NA	<0.5	
Gov. Office Bldg.	DT	44	<0.5	
School	DT	NA	<0.5	

* None detected.

all of these results were between 0.1 and 1.0 ppm. The bulk of these samples were collected at the four garment-manufacturing facilities, where the results of 181 samples ranged from 0.05 to 0.94 ppm, with most of these falling between 0.15 and 0.7 ppm[9-12]. Both personal and area samples were collected. At the specialty-chemical-manufacturing facility, the personal and area sampling results for eight samples ranged from 0.03 to 1.6 ppm[13], while at the scientific-glass-manufacturing facility, the results of three samples, both personal and area, ranged from 0.42 to 0.64[14]. Two area samples were collected in the area of the hospital where the oven for heating prosthetic materials was located; the detected formaldehyde concentrations were 0.37 and 0.73 ppm[15]. The results of 18 samples, both personal and area, collected at the paraformaldehyde-packaging plant ranged from less than 0.25 to 3.4 ppm. However, two short-term area samples were collected in locations which were not occupied by workers. If these results (2.0 and 3.4 ppm) are excluded, the highest level detected was only 1.91 ppm[16].

Of the seven facilities (six office buildings and a school) included in the low-exposure group, concentrations of formaldehyde were detected in two. The results of 39 samples ranged from 0.02 to 0.12 ppm, but almost all were below 0.1 ppm. Most of the results were in the vicinity of 0.05 ppm[17,18]. These results are summarized in Table 8[17-23].

DISCUSSION

As the data summary indicates, not every sample result falls into the concentration range of the exposure group in which the corresponding facility was placed. However, the vast majority of the samples are within the assigned ranges of less than 0.1 ppm for the low-exposure group, between 0.1 and 1.0 ppm for the intermediate-exposure group, and greater than 1.0 ppm for the high-exposure group. This trend supports the generalizations which were made regarding the placement of the facilities within the exposure groups. It is very important to realize that the generalizations which have been made regarding the facilities and the corresponding exposure ranges are not necessarily valid for any other facilities beyond those included in this presentation. It cannot be assumed for example, that all dialysis facilities will have exposure levels in the 0.2 to 0.7 ppm vicinity, or even in the 0.1 to 1.0 ppm range. Each facility must be dealt with based on the specific circumstances and conditions. The generalizations can be useful, however, in allowing selection of an appropriate sampling strategy when occupational exposures to formaldehyde are suspected or known.

Many of the industrial hygiene surveys which have been discussed were conducted in response to employer or employee requests, under the NIOSH Health Hazard Evaluation program. The others were part of the NIOSH Industry-wide Study on formaldehyde, which includes an epidemiologic study of workers exposed to formaldehyde in the garment-manufacturing industry. In the future, NIOSH plans to conduct several further industrial hygiene surveys in the garment-manufacturing industry in support of the epidemiologic study. NIOSH also plans to conduct surveys in other industries, to further characterize formaldehyde exposures. It is also anticipated that additional health hazard evaluation requests, concerning known or suspected formaldehyde exposures, will be received in the future, due to the continued interest in exposures to this compound.

REFERENCES

1. NIOSH, NIOSH Manual of Analytical Methods, Second Edition, Volume 1, P&CAM #125, April, 1977, DHEW (NIOSH), Washington, DC, Publication No. 77-157-A.

2. NIOSH, NIOSH Manual of Analytical Methods, Volume 6, P&CAM #318, August, 1980, DHHS (NIOSH), Washington, DC, Publication No. 80-125.

3. NIOSH, NIOSH Manual of Analytical Methods, Volume 7, P&CAM #354, August, 1981, DHHS (NIOSH), Washington, DC, Publication No. 82-100.

4. W. J. Chrostek, Health Hazard Evaluation Report No. HETA 81-142-892, June, 1981, DHHS (NIOSH), Cincinnati.

5. R. L. Ruhe, Health Hazard Evaluation Report No. TA 80-115-802, January, 1981, DHHS (NIOSH), Cincinnati.

6. P. L. Belanger, Health Hazard Evaluation Report No. HETA 81-211-984, October, 1981, DHHS (NIOSH), Cincinnati.

7. L. Elliott, Walk-Through Survey Report of the Dialysis Unit, Christ Hospital, Cincinnati, Ohio, Industrywide Study Report No. 125.18, February, 1982, DHHS (NIOSH), Cincinnati.

8. L. Elliott, Formaldehyde Concentrations During Animal Specimen Preparation,, NIOSH Internal Memorandum (Unpublished), dated June 7, 1982.

9. R. Keenlyside, L. Elliott, Health Hazard Evaluation Report No. HETA 81-056-854, April, 1981, DHHS (NIOSH), Cincinnati.

10. L. Elliott, Walk-Through Survey Report of the C. F. Hathaway Company, Waterville, Maine, Dover-Foxcraft, Maine, Industrywide Study Report No. 125.15, 1981, DHHS (NIOSH), Cincinnati.

11. L. Elliott, Walk-Through Survey Report of the Arrow Shirt Company, Lewistown, Pennsylvania, Industrywide Study Report No. 125.17, July, 1981, DHHS (NIOSH), Cincinnati.

12. L. M. Blade, L. T. Stayner, Walk-Through Survey Report of Manhattan Shirt Company, Americus, Georgia, Industrywide Study Report No. 125.22, 1982, (Unpublished Draft), DHHS (NIOSH), Cincinnati.

13. J. D. McGlothlin, P. Schulte, H. VanWagenen, Health Hazard Evaluation Report No. HETA 80-190-1135, June, 1982, DHHS (NIOSH), Cincinnati.

14. W. J. Chrostek, Health Hazard Evaluation Report No. HETA 81-098-941, August, 1981, DHHS (NIOSH), Cincinnati.

15. W. J. Chrostek, W. E. Shoemaker, Health Hazard Evaluation Report No. HETA 81-298-944, August, 1981, DHHS (NIOSH), Cincinnati.

16. L. Elliott, L. M. Blade, Walk-Through Survey Report of Vapor Products, Incorporated, Orlando, Florida, Industrywide Study Report No. 125.20, 1982 (Unpublished Draft), DHHS (NIOSH), Cincinnati.

17. A. G. Apol, Health Hazard Evaluation Report No. HHE 80-220-830, March, 1981, DHHS (NIOSH), Cincinnati.

18. W. J. Chrostek, Health Hazard Evaluation Report No. HETA 81-084-916, July, 1981, DHHS (NIOSH), Cincinnati.

19. P. L. Belanger, Health Hazard Evaluation Report No. HETA 81-164-982, October, 1981, DHHS (NIOSH), Cincinnati.

20. J. Horan, J. M. Bioano, Health Hazard Evaluation Report No. HETA 81-012-805, January, 1981, DHHS (NIOSH), Cincinnati.

21. S. A. Lee, Health Hazard Evaluation Report No. HETA 81-099-908, July, 1981, DHHS (NIOSH), Cincinnati.

22. NIOSH, Health Hazard Evaluation Report No. TA 81-007-985, 1981, DHHS (NIOSH), Morgantown, West Virginia.

23. N. Fannick, D. Baker, Health Hazard Evaluation Report No. HETA 81-161-952, September, 1981, DHHS (NIOSH), Cincinnati.

DISCUSSION

DR. DENNY: (Frank Denny, Veterans Administration.) Besides formaldehyde, glutaraldehyde is also used as a cold sterilization chemical. Does your method take into consideration, or can it distinguish, one aldehyde from another?

DR. BLADE: Which method are you referring to, the new NIOSH method?

DR. DENNY: Right.

MR. BLADE: The method is specific, and the oxazolidine that is formed between formaldehyde and benzylethanolamine is a specific one for formaldehyde. The method hasn't been tested, that I know of, to see if other aldehydes will give a reasonable result.

DR. DENNY: You don't know?

MR. BLADE: I don't know, but it wasn't designed for that purpose.

DR. DENNY: I have one other question -- You did not discuss the use of portable IR direct-reading instruments. Do you have any comment?

MR. BLADE: We don't use them, and that is upon recommendation of my laboratory support people. They have had some problems with that method.

MR. VAN HOUTEN: (Russell Van Houten, Liberty Mutual.) Your data didn't show any tests in building products manufacturing, such as plywood and particle board and such things. Have you made such tests, and do you contemplate them?

MR. BLADE: During the period of time that I intended to cover in this paper, we have not actually done any of those that I know of. This paper isn't all inclusive. There are probably some health hazard evaluations that we're not aware of, but I don't know of any. There were some tests done prior to 1980, for which I could probably get you results. Also, I mentioned that in the future we plan, after we have finished characterizing the exposed population for our epidemiology study, to look at other industries and continue characterizing exposures. Plywood and particle board are two of our priorities that we hope to look at some time. So, the answer to that part is yes.

MR. DU PONT: (Norman Du Pont, Hastings firm in Los Angeles.) You mentioned an epidemiological study of garment workers. What is the status of that study?

MR. BLADE: We have collected the personnel records on two of three facilities that we anticipate putting into the study. We are currently conducting walk-through industrial hygiene surveys combined with an inspection of personnel records at several other garment manufacturing facilities in hopes of locating a third suitable facility.

Our epidemologists' follow-up usually takes about two years. We anticipate collecting personnel records sometime during the third quarter or possibly the fourth quarter of this fiscal year. Our results would be available two years following that.

DR. BERNSTEIN: (Martin Bernstein, Ciba-Geigy.) Were all the samples that you gave results for done by the same method, or by all five methods?

MR. BLADE: They were done by all five methods. In the interest of trying to keep it understandable, I was forced to lump them all together.

DR. BERNSTEIN: Did the methods agree with each other within their limits of sensitivity?

MR. BLADE: In most cases, the methods continued to place the exposures within the ranges that were given. The only samples that I found that were greatly divergent were some of the samples that were collected using the charcoal tube method, and that was only in some cases.

In other words, on some of the surveys, even Method 318 or the charcoal tube method samples agreed, but there were a couple where they were greatly divergent; that was about the only real problem.

DR. BERNSTEIN: But, with the preferred method that NIOSH is using now, the last one, that limit of sensitivity was only

0.45 ppm, whereas a good percentage of your numbers were below that.

MR. BLADE: Yes, you see, first of all, not too many of the samples were taken with that method because that is rather new, and a lot of these surveys were done before that came out.

The lower limit of sensitivity is actually somewhat lower than that, so you can get results below 0.45 ppm, and they are probably valid. The usual procedure for our Division of Physical Sciences and Engineering is to test at approximately twice the standard, half the standard and maybe one-quarter the standard, or some such strategy. They have some sort of sampling strategy, but they did not go down to say, 0.1 ppm; but it seems to be valid. I have taken some samples since then, almost all of which were below 0.45 ppm, and they corresponded with the results of the CEA Instruments' direct reading instrument almost exactly. Those results were not included in this paper because I just got them last week. (Ed. Note: This breakdown is presented in Tables 7-15 of Mr. Blade's paper in this book.)

DR. BERNSTEIN: When this work is published, will there be a breakdown of what samples were taken with which instruments?

MR. BLADE: There wasn't intended to be. If you are interested, I can direct you to the references that were used for this paper, and you can look at all of them. (See Ed. Note above.)

DR. STERNBERG: (Stephen Sternberg, Memorial Cancer Center, Sloan-Kettering.) I would like to make a statement in regard to future epidemiologic studies, in particular in regard to the high exposure group, namely, personnel in gross histology laboratories. Glutaraldehyde has already been mentioned as another chemical that may confound the results. There are also a number of other fixatives used in those laboratories, such as mercuric chloride and picric acid and so on, so that results in terms of effects of exposure to formaldehyde in such a laboratory should be evaluated very carefully.

MR LARIME: (Michael Larime, Thetford Corporation.) Do you have any information on the ambient air (background) levels of formaldehyde in the locations that you measured and classified as low exposure facilities?

MR. BLADE: No.

MR. HELLINGS: (Clifford Hellings, American Cyanamid.) Question 1: What has been your experience with comparison tests for diffusion monitoring? Question 2: Are your plans for the future going to include this?

MR. BLADE: The results which I presented don't include the diffusional monitors for the reason that we don't find very good agreement in many cases between the diffusional monitors and the active sampling method. We have chosen active sampling. We are looking at diffusional monitors and we are doing some laboratory testing, too, but we are not sold on them yet. That is just our personal experience.

DR. VINCENT: (Frank Vincent, James River.) In this new method with benzylethanolamine, is the reagent available commercially?

MR. BLADE: Do you want the reagent or would you want completed tubes that are ready to sample?

DR. VINCENT: Completed tubes, if you have them.

MR. BLADE: They are available from Supelco, and they are called XAD-2 resin (formaldehyde tubes.)

DR. VINCENT: Thank you.

DR. DENNY: (Frank Denny, Veterans Adiministration.) I have one other question with regard to diffusion badges. Du Pont has one that incorporates colorimetric analysis; you referenced the CEA instrument, which is also a direct reading colorimetric monitor. Would you say that the correlation between the badge and the direct reading monitor would be valuable?

MR. BLADE: We have been finding that the data we have collected with that particular badge run consistently somewhat high. So, it doesn't actually correlate that well.

MR. WALTER: (Frank Walter, Manufactured Housing Institute.) One of the analytical procedures you mentioned was the P & CAM 125. The States of Wisconsin and Minnesota call for the use of that method in testing in new homes. Would you comment on the accuracy of that, what range for a 95 percent confidence, and is it a recommended procedure by NIOSH at this time?

MR. BLADE: NIOSH does not recommend that method specifically or approve methods, it only recommends them in general. However, the individual that I work with at the Division of Physical Science and Engineering has confidence in P & CAM 125. This is the fellow, by the way, that developed the new method, and he seems to have confidence in the impinger Method No. 125. I can't give you 95 percent confidence ranges. I think the published method itself has that information in it, but I can't say for sure.

DR. MIKSCH: (Robert Miksch, Air Quality Research.) I have one last comment on the diffusional samplers. They are personal samplers and are not really intended for area sampling. Did you notice any systematic difference in your sampling between your area sampling and your personal sampling?

MR. BLADE: With the diffusional monitors?

DR. MIKSCH: No, you did some personal sampling. Were they higher than the area samples?

DR. BLADE: With the same method, not really. You cannot always compare them but there was no trend going one way or the other. There was no noticeable trend, but many times an area monitor won't have the same exposure as a personal monitor for the obvious reason that it is not on the person.

MR. MC VEY: (Douglas McVey, Willamette Industries.) I have a question along the same line. In this presentation you put together your data for area monitors or tests along with personal testing. In the written presentation of the data, will that be broken down into two separate areas and distinguished or are they going to be lumped together?

MR. BLADE: Unfortunately, the time constraints did not allow me to do that today. It will not be separated in the Proceedings, but I can direct you to the references where you can look at all the samples instead of just the summmaries, and that would probably be the best way. (Ed. Note: See Tables 7-15 of Mr. Blade's paper in this book.)

MR. MC VEY: Is that liable to give us a biased number?

MR. BLADE: By combining them? There did not appear to be a trend that way.

CHAPTER 2

MATHEMATICAL CANCER RISK ASSESSMENT FOR FORMALDEHYDE

F. W. Carlborg

Consultant in Statistics
St. Charles, Illinois

Based on the experimental results from the CIIT experiment with rats exposed to formaldehyde, estimates of the risk of cancer at the low exposures of societal interest are made. Three representative mathematical dose-response functions (models) and their corresponding biological assumptions are considered in detail. Special attention is given to the assumption of low-dose linearity, which is the critical assumption in the most conservative method of risk assessment.

INTRODUCTION

The purpose of this paper is to present a mathematical cancer risk assessment for humans exposed to formaldehyde by inhalation based on the experimental results in rats as found in the experiment conducted by the Chemical Industry Institute of Toxicology (CIIT). Actually, risk assessments from several different points of view will be given, compared and commented upon. The fundamental assumption throughout is that the animals are surrogates for man, and no attempt will be made here to evaluate this assumption.

In the CIIT experiment, both sexes of Fischer 344 rats were used. About 240 rats were assigned to each of four dose groups, making about 960 rats in all. The target exposures were 0, 2, 6 and 15 parts per million concentration in the air, although the actual mean measured values were slightly different. The exposure began at about 1.5 months of life and lasted for 24 months. The laboratory exposure was intermittent: six hours a day for five days of the week. There were scheduled sacrifices for pre-selected animals at 6, 12, 18, 24 and 30 months after the first exposure.

For the ultimate purpose of a human risk assessment, it is convenient to convert each intermittent exposure level to its continuous lifetime equivalent, as if every breath contained this level. The target and continuous exposures are given in the first two rows of Table 1.

As it developed, the target organ was the nose, and the pathologic endpoint of interest (the response) was squamous cell carcinoma of the nasal turbinates. The earliest response was observed in a rat which died on the 358th day of the exposure. For each dose group, the number of rats in the group alive on that day and the ultimate number of responders are given at the bottom of Table 1.

One important human exposure to formaldehyde is that of industrial workers. Roughly 20,000 such persons are exposed to airborne concentrations of about 1.0 ppm while on the job. Their

TABLE 1

Relationship of Industrial Workers and Homeowner Exposure to Formaldehyde to the Exposure of Rats in the CIIT Study

Experimental Target Exposure (ppm)	0	2	6	15
Equivalent Continuous Exposure (ppm)	0	0.34	0.94	2.40
Industrial Worker (ppm)	0.13			
UF Homeowner (ppm)	0.004			
Number of Rats at Risk	216	218	214	199
Number of Responders	0	0	2	103

equivalent lifetime continuous exposure is 0.13 ppm, as shown in the third row of Table 1.

A second important human group is those persons exposed as a result of living in a home insulated with urea-formaldehyde (UF) foam. Based on a report by the Consumer Product Safety Commission (CPSC), the worst-case exposure has a continuous lifetime equivalent of 0.004 ppm, as shown in the fourth row of Table 1.

Thus, the primary goal for this paper is to use the animal results given in Table 1 to estimate the risk for each of the two human situations given in Table 1.

METHODS, RESULTS AND DISCUSSION

To position the various methods of risk assessment in an introductory way, consider the two questions in the margins of Table 2: (1) Is there a universal threshold in the dose for the entire population such that the risk below this threshold

TABLE 2

Relationship Between the Ideas of a Universal Threshold for Effect and the Shape of the Dose-Response in Decision-Making

Presence of an Universal Threshold	Consideration of the Shape of Dose-Response	
	Response	
	NO	YES
	DECISION	
YES	NOEL	Steep Curve, Large Threshold
NO	Forced Low-Dose Linearity	Steep Curve, Small Risk

is zero? (2) Does the risk at a low dose (or the threshold) depend on the shape of the dose-response curve at the higher doses?

A "yes-no" answer to these two questions, respectively, leads one to the traditional method involving the no-observed-effect level (NOEL), which is usually interpreted as a lower bound on the universal threshold.

A "yes-yes" answer leads to a model which provides a numerical estimate of the threshold, using all the observed data and reflecting the steepness of the dose-response curve.

In effect, regulatory agencies in the USA in recent years have chosen a "no-no" answer and have chosen low-dose linearity for risk assessment. With this method the shape of the dose-response curve has essentially no effect on the ultimate estimated risk.

A "no-yes" answer leads to one of the methods sensitive to the shape of the dose-response curve. When the curve is steep, these methods estimate a small risk at a low dose. When the curve is shallow, these methods estimate a large risk at a low dose. Four of the models in this class are the probit, the logit, the multi-hit (gamma), and the Weibull model. With a steep dose-response curve like that for formaldehyde, the differences among these four models are small. The Weibull model is generally the most conservative. For this and other reasons, it will be taken here as the representative of this class.

Mathematical definitions of particular models are as follows:

GENERAL FORM

$$P(d) = 1 - e^{-f(d)},$$

Where:
P = Probability of a response
d = The dose
f = A function of the dose

UNIVERSAL THRESHOLD

$$f(d) = 0 \quad \text{FOR } d < \gamma$$
$$= B(d - \gamma) \quad \text{FOR } \gamma < d$$

WEIBULL

$$f(d) = Bd^m$$

POLYNOMIAL (MULTISTAGE)

$$f(d) = B_1 d + B_2 d^2 + \ldots + B_m d^m \quad \text{FOR } B_i \geq 0$$

First, there is the general form with the function f of the dose in the exponent. Each of the three models is then a special case of this general form.

The universal threshold model is associated with the upper-right box of Table 2. The parameter (γ) is the threshold.

The Weibull model is associated with the lower-right box of Table 2. The parameter m is the important shape parameter, which will be discussed again later.

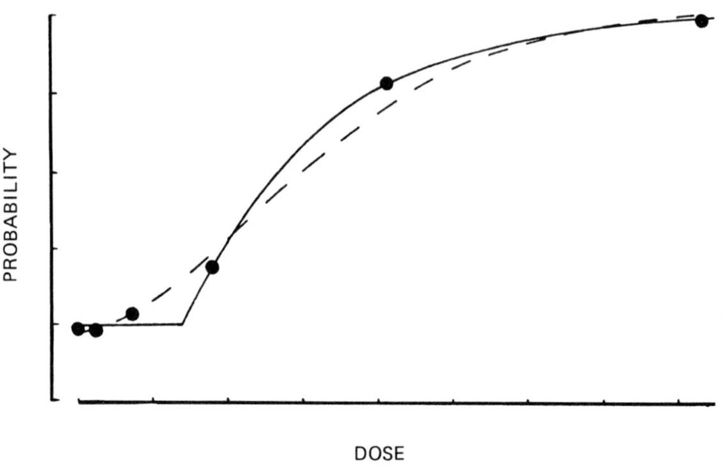

Fig. 1. An illustration of the Universal Threshold Model (solid line) and the Weibull Model (dashed line) as fit to the same data set (not formaldehyde).

As an illustration, Fig. 1 shows the universal threshold model (the solid curve) and the Weibull model (the dashed curve) successfully fitted to the same data set -- a data set not related to formaldehyde.

One way of obtaining low-dose linearity is through the polynomial model defined norm. When the parameter B_1 in the linear term is not zero, then low-dose linearity exists, and the model is associated with the lower-left box of Table 2.

The numerical results of fitting the models to the experimental results are given in Table 3. First notice that four versions of the polynomial model have been used. For example, the "1.5" refers to a version of the polynomial model containing only a linear and fifth order term (B_1d and B_5d^5).

Second, consider the calculated numbers of responders for each model or version. Recall that the observed numbers of

TABLE 3

Risk to Man from Formaldehyde Exposure as Calculated from Observations in Rats (CIIT Study)

Model	Calculated Number of Responders at Dose (ppm)				Risk (Per 10^6)	
	0	2	6	15	Worker	UF Home
Universal Threshold	0	0.0	2.0	103.0	--	--
Weibull	0	0.0	2.0	103.0	0.9	0.01
95% CL					10.	0.01
Polynomial						
*1,5	0	0.2	1.9	103.1	310.	10.
2,5	0	0.1	2.0	103.0	53.	0.1
3,5	0	0.0	2.0	103.0	8.	0.01
4,5	0	0.0	2.0	103.0	1.	0.01
*95% CL from CPSC					1600.	51.

responders were 0, 0, 2 and 103, respectively. Notice that each of the models provides an excellent fit to the observed results. Thus, it is clear that statistics cannot determine the correct answers to the two basic questions of Table 2.

Third, the polynomial or multi-stage model has a basic defect which is illustrated in the bottom half of Table 3. How does one choose that best fit among these four versions (plus others not shown) which fit the observed data extremely well and essentially equally well in the statistical sense?

Fourth, consider the risk to the industrial worker having an equivalent lifetime continuous exposure of 0.13 ppm. The universal threshold model provides a best estimate of 0.92 ppm for the threshold, indicating that the worker's exposure is about one order of magnitude below this threshold. As shown in the next-to-the-last column of Table 3, the best estimate of the risk to this worker according to the Weibull model is 0.9 in a million. The corresponding one-sided upper 95% confidence limit is 10 in a million. The best estimates under the indicated versions of the polynomial model range from 1 in a million to 310 in a million. The Consumer Product Safety Commission, working with a slightly different preliminary report of the experimental results, produced a risk assessment using the polynomial model. Their algorithm actually selected a version without the linear term as a best estimate. On the assumption that the linear term should be there, it was introduced by way of an upper confidence limit. Their resulting version applied to the industrial worker produces an estimated risk of 1600 in a million as shown at the bottom of Table 3.

Finally, the last column of Table 3 gives the corresponding estimated risks for the person living in a UF-insulated home.

Contrary to the impression created in Table 3, the basic problem for mathematical risk assessment is not which dose-response model to select, but rather how to answer the two basic questions of Table 2: the first concerning a universal threshold and the second concerning the importance of the shape of the dose-response curve. The answers given to the these two biological questions largely solve the problem of model selection. The remainder of this paper reconsiders these two questions from a statistician's point of view.

For motivational purposes, imagine a population of animals exposed to a carcinogen over a normal lifetime. At a given lifetime dose, some of the animals will respond (develop a tumor), and some will not. At a higher dose, more will respond; at a lower dose, fewer will respond. This implies that each animal has its individual threshold with respect to the dose. Over the entire population of animals (not just those in the experiment), there is some distribution of individual thresholds.

The existence of a universal threshold, say γ, in this context means that all the individual thresholds lie above γ. The implied risk is zero at any exposure below γ. The practical problem with a given data set is merely to estimate the universal threshold by one of the two methods mentioned in Table 2.

The absence of a universal threshold means that for any small exposure there is some proportion of animals who will respond, because their individual thresholds lie below this small level of exposure. The problem is to estimate this proportion. One way to do it is to assume that the distribution of individual animal thresholds has some standard statistical form, to estimate the parameters of this distribution, and then to calculate the desired proportion from these estimated parameters.

The Weibull distribution has been chosen here as the representative of a class of possible standard distributions. Within the context of the Weibull model, there is wide range of possibilities, with a correspondingly wide range of low-risk assessments. Toward one extreme, the distribution of individual thresholds may be tight, as indicated by the short-dashed curve in Figure 2. Then the dose-response curve is steep, as indicated by the short-dashed curve in Figure 3. Also, the value of the shape parameter m is then large, and the calculated risk at a low dose is then small.

Toward the other extreme, the distribution of the individual animal thresholds may be flat, as indicated by the solid curve in Figure 2. Then the dose-response curve is shallow, as indicated by the solid curve in Figure 3. Also, the value of the shape parameter m is then small, and the calculated risk at a low dose is then large.

Intermediate situations are possible, as indicated in Figures 2 and 3. With actual reported data sets, the value of m ranges from about 0.5 to about 6.5. A value of m = 1 indicates low-dose linearity. For formaldehyde, the estimated value of m is 4.6, making it the second largest ever encountered by the author.

The basic question for risk assessment is: Should this observed information about the distribution of the individual thresholds have an important effect on the ultimate low-risk assessment? A "yes" answer leads to the Weibull model or something similar to it.

Regulatory agencies in the USA in recent years answer "no" to this question. They assume that there is low-dose linearity of the dose-response curve regardless of what the experimental

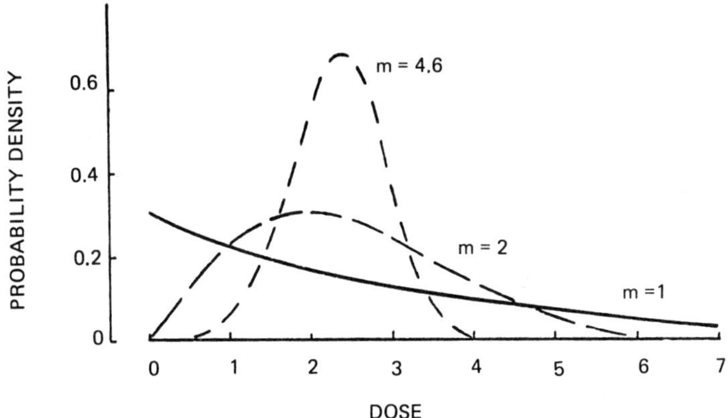

Fig. 2. The distribution of individual animal thresholds utilizing various shape parameters.

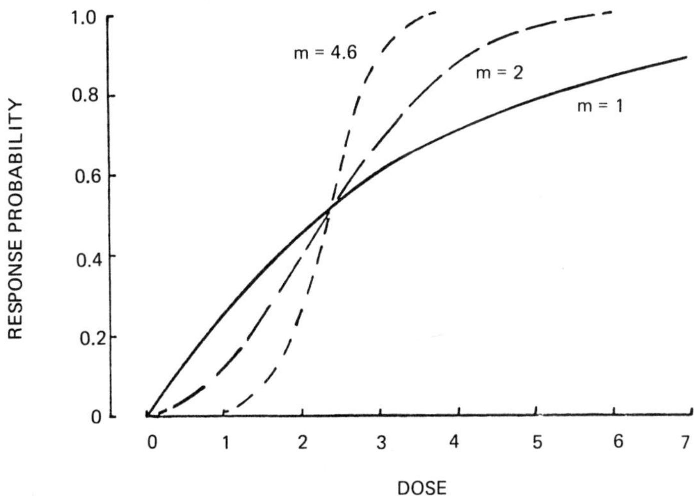

Fig. 3. The effect of shape parameter size on the shape of the dose-response curve.

results say about the distribution of the individual thresholds. In the language of the Weibull model this becomes: At low doses, use a value of $m = 1$ for the shape parameter, regardless of the observed value of the parameter m.

It is important to observe that this is a policy decision, not based on a scientific hypothesis. A scientific hypothesis must be testable, rejectable, or falsifiable by experimentation. No scientific experiment can reject the hypothesis of low-dose linearity. Regardless of the experimental outcome, one can always claim that farther down in the doses beyond the practical experimental limit there is linearity.

Regulatory agencies frequently cite a mathematical proof of low-dose linearity as follows:

GENERAL FORM: $P = 1 - e^{-f(d)}$

PROOF OF LOW-DOSE LINEARITY

ASSUMPTIONS

$f(d) = f(d_0 + cd)$

d = APPLIED (EXTERNAL) DOSE OF THE CARCINOGEN.

c = A CONSTANT OF PROPORTIONALITY.

d_0 = CARCINOGENICALLY EQUIVALENT DOSE OF OTHERS AT THE SITE.

TAYLOR SERIES EXPANSION:

$F(d) = B_0 + B_1 d + B_2 d^2$ + HIGHER ORDER TERMS IF NEEDED

B's DEPEND ON f, d_0 and c

$B_1 = 0$ IF AND ONLY IF $f'(d_0) = 0$

FURTHER ASSUMPTION: $f'(d_0) = 0$

Like all mathematical proofs, there are assumptions. The first is that the species of primary importance (man) already has an incidence of cancer at the target organ of interest. Presumably, this is caused by the presence of other carcinogens to which the one in question, say formaldehyde, is to be added. The second is that for any applied (external) dose d of formaldehyde the effective (internal) dose of formaldehyde at the target site is proportional to the applied dose, as in cd. Under these two assumptions and one more, low-dose linearity of the dose-response curve is proved. If either of these two assumptions is false, then the proof does not hold. In fact, one can derive the Weibull model at low doses just by changing the second assumption to read that the effective dose is proportional to the applied dose raised to a power, as in cd^m.

The Consumer Product Safety Commission says that formaldehyde is genotoxic and by implication says that this establishes the additivity of doses. Mathematically, however, this is not sufficient unless the genotoxic mechanism of formaldehyde is identical to the genotoxic mechanism of the other carcinogens to which it adds at the site.

The two assumptions for this proof are as experimentally irrefutable as the outright hypothesis of low-dose linearity. For example, to disprove dose-wise additivity it would be necessary to test formaldehyde for additivity with respect to all human carcinogens at all potential sites in the human body. This is clearly impossible. Thus, the mathematical proof of low-dose linearity depends on at least two assumptions which are scientifically empty, because they cannot be challenged experimentally.

In summary, the forced use of low-dose linearity for risk assessment in the case of formaldehyde is strictly a policy decision and not a scientific necessity.

DISCUSSION

MS. MARGOSCHES: (Elizabeth Margosches, Environmental Protection Agency.) I am curious whether you feel that the interim sacrifices may have made a difference in being able to use those data for modeling in the way you did, and also, whether you have incorporated time in any way; not necessarily a time-to-tumor model? For instance, I have tried a Weibull form with a lag for ability to see the tumor.

DR. CARLBORG: I used the interim sacrifices and did the life table analysis, and then fit the Weibull model and other models to those data. I obtained essentially the same results that I obtained by just taking all the animals alive at the first tumor. The differences between doing it that way and the way I did here are so small compared to the other differences that I was talking about, I did not think it was worth bringing up.

The second question is about the time-to-tumor information which is available because there were serial sacrifices, and I have looked at that. Really the only place you can get anything out of that is in the high-dose group where there were a large number of responders. As I recall, there were only two in the second highest dose, and the time-to-tumor curve fits a Weibull model with that as a variable. The shape parameter for that comes out to be about 6 1/2 which is kind of a standard one for carcinogens, at least in my experience and my reading.

DR. LAMM: (Dr. Steven Lamm, CEOH.) Your argument for the demonstration of a universal threshold for formaldehyde was based on a Weibull analysis. Your conclusion from it then became that, as a result of that demonstration, a Weibull analysis was appropriate for the mortality data or for the incidence data of nasal carcinomas. Is this a bit of circularity in using the Weibull model in order to conclude that the Weibull model in the next step in justifiable? If you used a different model on your original question, would your conclusion have been otherwise?

DR. CARLBORG: No. I don't think I made the point very clearly. You recall Fig. 2, I said there were two questions leading to four boxes. When a person doing the risk assessments decides the answer to those two questions, then a mathematical model can be provided which will fit the data and will produce an answer which is consistent with his answers to those two questions.

I did not use and did not attempt to use or intend to use the Weibull model to prove that there was a universal threshold. The Weibull model fundamentally assumes there is no universal threshold and attempts to do low-dose risk assessment under that assumption. So, I don't think that the circularity which you mentioned is really there.

CHAPTER 3

CASE CONTROL STUDY OF CANCER DEATHS IN DU PONT WORKERS WITH POTENTIAL EXPOSURE TO FORMALDEHYDE

W. E. Fayerweather, S. Pell and J. R. Bender

E.I. du Pont de Nemours & Company, Inc.
Wilmington, Delaware

The objective of this study was to determine whether occupational exposure to formaldehyde increases Du Pont workers' risk of developing cancer. Special interest centered around cancers of the lung and upper respiratory tract, particularly the nasal cavities.

A case-control study of cancer mortality among workers exposed to formaldehyde was completed. Cancer deaths from 1957 through 1979 were studied at eight formaldehyde manufacturing or using plants. There were 481 cancer deaths at these sites, 142 of which were among workers with potential exposure to formaldehyde. The data were analyzed by tumor site, latent period, duration of exposure, exposure level and frequency, cumulative exposure index, age and year of death, and age and year of first exposure. Analyses of lung cancer deaths were adjusted for subjects' cigarette smoking habits.

In none of these analyses was the formaldehyde workers' relative risk of cancer significantly greater than 1.0 ($p > 0.05$). There were no nasal cancer deaths and no lung cancer excesses among workers exposed to formaldehyde. These results were strengthened considerably by factoring workers' smoking habits into the analyses. Slight, but statistically nonsignificant elevations in relative risks were observed for prostatic and bladder cancers. These elevations showed no indications of being causally related to formaldehyde and appeared to be due to chance.

In summary, a thorough analysis of the data suggested that cancer mortality rates in Du Pont's formaldehyde-exposed workers were no higher than the rates among non-exposed co-workers. These results provide further support for the Du Pont Company's position that airborne levels of 1 ppm time-weighted average and a 2 ppm ceiling provide adequate worker protection.

INTRODUCTION

The study's objective was to determine whether occupational exposure to formaldehyde increases the Du Pont workers' risk of dying from cancer. Special interest was to center around cancers of the lung and upper respiratory tract, particularly the nasal cavities.

The present study was initiated in 1980 after receipt of the preliminary results of a study by the Chemical Industry Institute of Toxicology (CIIT). The CIIT study was entitled: "A Chronic Toxicology Study in Rats and Mice Exposed to Formaldehyde."[1] According to the final CIIT report, rats and mice were exposed to 0, 2, 6, and 15 ppm of formaldehyde gas for six hours per day, 5 days per week for twenty-four months. At 15 ppm about 50% of the rats and 1% of the mice developed nasal cancer, primarily squamous cell carcinoma. One percent of the rats also developed nasal cancer at 6 ppm, but this number was not statistically significant ($p > 0.05$). No nasal cancers were seen at 2 ppm or at 0 ppm.[2]

The carcinogenic response observed in the CIIT study was highly dependent on species, exposure level, and target tissue. Squamous cell carcinomas were observed only at exposure levels that were extremely irritating to the animals and that produced extensive cell destruction. This cell destruction was followed by rapid cell proliferation, increases in cell turnover rates, and squamous cell metaplasia. Metaplasia regressed in animals maintained 3 months after the 24-month exposure period.[2]

A Biodynamics chronic inhalation study exposed rats, monkeys, and hamsters to 3, 1, and 0.2 ppm formaldehyde for 22 hours per day for 6 months. No tissue damage was observed at 1 ppm or below in rats and monkeys.[2]

Three epidemiologic studies specifically designed to assess the cancer risk from formaldehyde exposure in humans have shown no excesses of cancer that could be attributed to formaldehyde. These studies include the Walrath[3] proportionate mortality study of New York morticians[3], the Marsh[4] proportionate mortality study of formaldehyde resin plant workers[4], and the Wong cohort mortality study of formaldehyde manufacturing plant worker[5].

Although there were no nasal cancer deaths observed in these three epidemiologic studies, some of the authors reported slight elevations in risks for certain other sites. The Walrath[3] study found slight excesses for cancers of the skin, kidney and brain. Wong[3] found slight excesses for prostatic and lymphopoietic cancers. Marsh[4] on the other hand reported that rates for formaldehyde workers were "generally similar to those for unexposed workers."

When the results of the epidemiologic studies are viewed as a whole, the data suggest that formaldehyde has not been responsible for producing cancer in man. However, Wong's study was limited by small numbers. The proportionate mortality (PMR) studies of Marsh and Walrath were limited by the PMR design itself. Therefore, to help clarify the question concerning formaldehyde's effect in humans, Medical Division decided to conduct its own epidemiologic study of cancer risk among Du Pont formaldehyde workers.

METHODS

General Design

This was a matched-pairs case-control study of all male cancer deaths that occurred among active and pensioned employees at eight study plants from 1957 through 1979. For each cancer death, a control matched on age, payclass, sex, and adjusted service date (ASD) was chosen. Work histories and smoking histories were sought for all cases and controls. Jobs were categorized by frequency and intensity of exposure to formaldehyde. Estimates of relative risk (or odds ratios) were computed as measures of the cancer risk in formaldehyde-exposed workers relative to the risk in non-exposed workers.

Study Populations

1. Plant Selection

There were several criteria for including a plant in the study:

There must have been at least 15 years since formaldehyde was first used at the plant.

The plant's use of formaldehyde must have been on a continuous basis.

Recent exposure levels must have occasionally been at the odor threshold or above.

The plant must have had a stable population.

The following plants were chosen to participate in the study:

Plant (Dept.)	Years Formaldehyde was Present
Parlin (F&F)	1940-1972
Belle (BIOCHEM)	1939-present
Parlin (PHOTO)	1960-present
Perth Amboy (ELECTROCH)	1920-1970
Rochester (PHOTO)	1945-present
Toledo (C&P)	1947-present
Toledo (F&F)	1945-present
Washington Works (PPD)	1960-present

2. Definition and Selection of Cases

Cases were defined as all male cancer deaths that occurred among active and pensioned employees at the study plants during the period 1957-1979. Only those cases for whom work histories could be ascertained were included in the study. Females were excluded from the study because the formaldehyde exposed work force was primarily male.

3. Definition and Selection of Controls

In general, controls were selected from the cases' co-workers who were at risk at the time of the onset of the cases. A design in which controls are selected from the population at risk has been shown to be superior to the classical design in which controls are selected from persons unaffected as of a certain date[6].

Each cancer case was matched with one male employee who was on the company's active rolls during the case's last year of employment. To be eligible for selection, the control had to match the case's age ($\pm$3 years), adjusted service date ($\pm$3 years), plant location (exact), sex (males only), and payclass (wage or salary) as of the case's last year at the plant. For any given case the control was the first eligible employee appearing on the payroll roster after the case's name. (adjusted service date is similar to the employee's hire date, only it has been adjusted to reflect leaves of absence and lay-offs.)

4. Definition of Exposure

Within each plant, jobs were divided into three exposure categories: continuous-direct, intermittent, and background. These categories were based on both the frequency and intensity of exposure without regard to the use of respirators. Since routine use of respirators began only within the past 2 to 5 years, respirator usage did not complicate the exposure estimates.

Continuous-Direct Exposure

These were jobs in which a worker was assigned to a discrete area in which formaldehyde was produced, separated, recovered, processed, or loaded/unloaded. Exposure to formaldehyde must have occurred at least three days out of five to be included in this group. Within this group, air exposures were ranked according to their approximate eight-hour time-weighted average concentrations:

Level 1 = less than 0.1 ppm

Level 2 = from 0.1 ppm up to but not including 2.0 ppm

Level 3 = 2.0 ppm or more

The actual ppm values associated with each level were extrapolated from air monitoring data whenever possible. When air monitoring data were unavailable, long-service employees were able to recall certain acute effects associated with formaldehyde exposure, such as odor or sensory irritation (eye, nose, or throat).

Intermittent Exposure

These were jobs in which a worker was not assigned to a discrete area where formaldehyde was produced, separated, recovered, processed, or loaded/unloaded. However, the job required that the worker periodically work in these areas. These jobs were further divided into two exposure categories:

Low - jobs having little potential for permitting peak exposures to formaldehyde of the order of 2 ppm.

High - jobs permitting a worker to be exposed to peak formaldehyde concentrations of the order of 2 ppm or more.

The intermittent category typically included maintenance personnel assigned on a plant-wide basis, and laboratory quality control people who supported non-formaldehyde as well as formaldehyde units. The formaldehyde-exposed workers not falling into the continuous-direct category were placed into the intermittent category.

Background Exposure (Nonexposed Group)

These were all jobs that did not fall into the continuous-direct or intermittent groups.

Sources of Data

1. Cancer Deaths (Cases)

All deaths among active and pensioned employees were identified through the Company's Mortality File. This file was begun in 1957 and was current at the time of this analysis through December, 1979. The Epidemiology Section's trained nosologists code each cause of death according to the ICD revision in effect at the time of death. The Company's noncontributory group life insurance plan is the mechanism by which deaths are reported in the mortality file.

2. Controls

Eligible controls were selected from annual plant payroll rosters that are kept on file in Wilmington Medical Division. These annual payroll rosters include all active employees at a plant during a given year. Wage roll employees appear on separate lists from salary roll employees. The lists are offered by location and department, either alphabetically or by birth year. These lists contain the co-worker population at risk as of the case's last year at the plant.

3. Work Histories

Work histories on all cases and controls were supplied by the plants. These histories for the large majority of subjects were obtained directly from personnel records. When personnel records were unavailable, medical records and interviews with co-workers were used to ascertain the work histories. When work history data were ascertained through interviews, long-service co-workers who had knowledge of both the worker and of the jobs with potential formaldehyde exposure were consulted.

The work history data that were collected on each case and control included:

Name

Social Security Number

Sex

Birth date

Date hired

Payclass when hired

Final payclass

Date of final departure from the plant

Reason for departure

All jobs held at the plant, including dates in and out of these jobs

Work history prior to working at Du Pont, if available

The formaldehyde exposure potential for each of the Du Pont jobs held.

This information was abstracted from plant personnel records, transcribed to code sheets (Appendix A), and forwarded to Wilmington Medical Division.

4. Assessment of Formaldehyde Exposure Potential

Prior to assessing exposure potentials, the plant's industrial hygienists and other similarly qualified personnel were asked to complete three items: process descriptions, job code dictionaries and exposure potential dictionaries.

- Process descriptions included descriptions of the site's present and past processes, work areas, and jobs with potential formaldehyde exposures.

- Job code dictionary: Formaldehyde-associated work areas and jobs were defined, tabulated, and assigned numeric codes.

- Exposure potential dictionary: Jobs by work area and time period were assigned to the exposure categories of continuous-direct, intermittent, or background.
- Continuous-direct jobs were further subdivided into levels 1-3, and intermittent jobs were subdivided into levels 1-2 (see exposure definitions given above).

Exposure potentials were extrapolations based on:

- Recent air monitoring data for current jobs and work areas. The Company's current airborne formaldehyde standard is a 1 ppm 8-hour TWA with a 2 ppm ceiling.

- Past air monitoring data from 5-7 years ago.

- Statements by long-service employees as to whether and how often formaldehyde was present in high enough concentrations to produce an odor or sensory irritation.

- Knowledge of odor and sensory irritation thresholds. The thresholds for odor recognition and sensory irritation vary considerably among individuals. The former has been reported at concentrations between 0.1 and 1 ppm, while sensory irritation is generally not reported until concentrations approach 0.5 to 1 ppm. As concentrations rise above 2 to 5 ppm, there will be definite complaints of eye, nose, and throat irritations.

- Knowledge of past process changes and of engineering or personal controls.

5. Smoking Histories

Smoking histories of cases and controls were ascertained primarily by proxy, that is by interviewing subjects' living co-workers. In a few instances living controls were either interviewed directly, or smoking habits were ascertained from existing medical records.

All proxies were long-service active or pensioned employees who had either supervised the subject, worked in the same area, or been a close friend of the subject.

By reviewing personnel records and by questioning other long-service employees, plant personnel determined which co-workers were best suited to serve as proxies.

The proxy-completed questionnaires (Appendix B) allowed the cases and controls to be ranked according to the number of packs of cigarettes smoked per day (none, less than 1/2, 1/2-1, 2 or more packs/day). Completed smoking history questionnaires were forwarded to Wilmington Medical Division for processing.

Data Processing and Quality Control

All smoking and work histories were abstracted from plant data sources and were transcribed to code sheets by plant personnel. These code sheets were then forwarded to Wilmington Medical Division for processing. Each plant also forwarded personnel records for several cases and controls. These records were checked against the completed code sheets to verify that the coding was accurate and complete.

The smoking and work history data were entered directly from code sheets into computer files. Computer routines were written to check for missing or invalid entries and to edit the data base. For all cases and controls the computer's version of the data was checked against the original code sheets for accuracy.

Data Analysis

In this study the relative odds ratio was used as the measure of association between exposure and disease. The relative odds ratio is the ratio of the odds of a case having been exposed relative to the odds of a control having been exposed. An odds ratio of 1.0 indicates no association.

When controls are selected from the population that gave rise to the cases, the relative odds ratio closely approximates the relative risk ratio[6], another measure of association between exposure and disease. The relative risk ratio is the ratio of the exposed group's risk of dying from cancer relative to the nonexposed group's risk of dying from cancer. The terms relative risk ratio and relative odds ratio have been used interchangeably in this study.

Adjusted (or matched-pairs) odds ratios, lower and upper 95% confidence limits on the odds ratios, and two-tailed p-values were computed for all analyses. These adjusted odds ratios were computed in such a way as to take into account the matched-pairs design. The analyses of lung cancer also were adjusted for between-group differences in smoking habits. The adjustments for matching and smoking habits were done by the conditional logistic regression method of Breslow and Day.[7]

The nonmatched (crude) odds ratios were similar in direction and magnitude to the adjusted (matched) odds ratios. Therefore, for descriptive purposes and ease of understanding, the crude odds ratios were presented for most analyses. However, statistical significance tests were based on the appropriate matched analyses.

The data have been analyzed by:

Vital status of the control

Source of work history

Payclass

Tumor site (major organs and systems)

Latency

Duration

Frequency of exposure (intermittent vs continuous)

Exposure level

Cumulative exposure index

Year of death

Age at death

Age at first exposure

Year at first exposure

Smoking habits

Plant site

Each cancer site was analyzed once for each of six different hypothetical latent periods: 0-4, 5-14, 15-24, 15+, 20+, and 25+ years. In subsequent analyses only latent periods that maximized the odds ratio in these initial analyses were used. Such an approach tends to minimize exposure misclassification that results from misspecification of the latent period.[8] However, it also tends to slightly bias the analyses towards over-estimating the true odds ratio.

Latency and exposure were treated in the following manner. A history of formaldehyde exposure was counted only if it occurred during a certain time period prior to the case's death. For instance, when a latent period of 20+ years was assumed, only exposure that occurred 20 or more years prior to the case's death was counted. This means that for controls, only exposure that occurred 20 or more years prior to the matching case's death was counted. Furthermore, formaldehyde exposure that occurred in a control after the matching case's death was ignored.

In all instances two-tailed significance tests were performed, and significance was judged at the 0.05 probability level.

RESULTS

Demographic Description of Cases and Controls

From 1957 through 1979 there were 493 cancer deaths associated with the eight study plants (Table 1). Of these 493 cases, 12 (2.5%) had to be excluded because work histories were unavailable. The twelve excluded cases consisted of 6 lung, 4 digestive, 1 lymphopoietic, and 1 other/unspecified cancer death. For about 2% of the cases, a second control had to be chosen because work histories were unavailable for the first control chosen. Eighteen percent (86) of the 481 cases were on the active rolls and eighty-two percent (395) were pensioners at the time of death.

The distribution of cancer deaths by tumor site was not unusual (Table 2). The tumor sites with the largest number of cases were lung (181), colon and rectum (54), lymphopoietic (49), prostate (33), and stomach (23). There was only 1 case in the category of nose, nasal cavities, eustachian tube and middle ear. This case was actually a eustachian tube/middle ear cancer, not a nasal cancer.

TABLE 1

Numbers of Cancer Deaths in Each Plant

Plant Site	Cancer Deaths 1957–79	Cancer Deaths 1957–79 That were Included in The Study*
Parlin F&F	87	79**
Belle	220	218+
Parlin Photo	83	83
Perth Amboy	10	10
Rochester Photo	35	35
Toledo C&P	3	3
Toledo F&F	19	19
Washington Works	36	34++
Total	493	481

* A total of twelve cases were dropped from the study because work histories were unavailable for these cases.

**Work histories could not be determined for 4 lung cancer deaths, 3 digestive cancer deaths, and for 1 other/unspecified cancer death.

+ Work histories could not be determined for 1 lung cancer death and for 1 digestive cancer death.

++Work histories could not be determined for 1 lung cancer death and for 1 lymphopoietic cancer death.

The distribution of cancer deaths by age group was about as expected. Lung, colon and rectum, lymphopoietic, and stomach were most common in the 60-69 year age group, whereas prostate was most common in the 70-79 year age group. Because an age-matched control was selected from each case, the average ages of cases (65.1 years) and controls (64.4 years) were very similar.

The ratio of wage to salary cases was about 3 to 1, which corresponds roughly to the actual wage/salary ratio for the plant populations.

The number of cases increased steadily from 98 in the period 1957-1964 to 142 in the period 1975-1979 (Table 3). This increase reflects a shift in the age distribution and an increase in the number of active and retired employees.

The prevalence of cigarette smoking was higher among cases than among controls. The all-cancer adjusted odds ratio of workers who smoked 2 or more packs of cigarettes per day were unknown or unreported for about 10% of cases and controls. About 22% of the cases and 38% of the controls were nonsmokers.

With respect to lung cancer, only 12% of the cases were nonsmokers compared to 36% of controls. The risk of lung cancer in smokers relative to nonsmokers showed a strong dose-response relationship. The adjusted odds ratio increased from 2.1 in those who smoked less than 1/2 pack per day to 10.3 in those who smoked 2 or more packs per day (Table 4).

Odds Ratios by Latent Period

The adjusted odds ratios generally showed little variation from one latent period to another (Table 5). The one exception

TABLE 2

Distribution of Cancer Deaths by Plant Site and Tumor Site

Tumor Site	Parlin (F&F)	Parlin (Photo)	Toledo (F&F)	Toledo (C&P)	Perth (Amboy)	Rochester (Photo)	Belle (Bioch)	Washington Works (PPD)	Total
-Buccal Cavity, Pharynx	2	1	0	0	0	1	2	1	7
-Esophagus	1	0	1	0	0	3	4	0	9
-Stomach	7	7	0	0	3	2	4	0	23
-Small Intestine	1	0	0	0	0	0	0	0	1
-Colon and Rectum	12	13	1	0	0	4	20	4	54
-Liver, Gall Bladder	1	2	0	0	0	0	5	2	10
-Pancreas	5	7	0	0	0	0	4	2	18
-Eustachian Tube/Middle Ear	0	0	0	0	0	0	0	1	1
-Larynx	3	0	1	0	0	1	3	0	8
-Lung, Bronchus	19	20	10	2	4	10	104	12	181
-Other Respiratory	0	1	0	0	1	0	0	0	2
-Pleura	1	0	0	0	0	0	0	0	1
-Prostate	4	5	2	0	1	2	19	0	33
-Testis	1	1	0	0	0	0	0	1	3
-Other Genital	1	1	0	0	0	0	1	0	3

-Kidney	0	3	0	0	0	1	9	0	13
-Bladder	2	2	1	0	0	1	4	0	10
-Lymphopoietic	5	11	2	0	0	5	22	4	49
-Brain, Other Nervous	2	4	0	0	0	1	4	1	12
-Connective Tissue	1	1	0	0	0	0	0	0	2
-Malignant Melanoma	2	1	0	0	0	0	3	1	7
-Other Skin	0	0	0	0	0	0	1	0	1
-Other Endocrine	0	0	0	0	0	0	0	1	1
-Bone	3	1	0	0	0	1	0	2	7
-All other Ill-Defined/ Unspecified	6	1	1	1	1	3	9	2	24
-Peritoneum, Retroperit.	0	1	0	0	0	0	0	0	1
-All sites	79	83	19	3	10	35	218	34	481

TABLE 3

Distribution of Cancer Deaths by Year of Death and by Tumor Site

Cause of Death by Tumor Site	1957-64	1965-69	1970-74	1975-79
Buccal Cavity, Pharynx	2	3	1	1
Esophagus	1	1	3	4
Stomach	7	4	7	5
Small Intestine	0	0	1	0
Colon and Rectum	13	8	17	16
Liver, Gall Bladder	2	1	5	2
Pancreas	7	1	3	7
Peritoneum	0	0	0	1
Eustachian Tube, Middle Ear	1	0	0	0
Larynx	0	4	2	2
Lung, Bronchus, Trachea	27	46	50	58
Other/Ill-Defined Resp.	0	1	0	1
Pleura	0	0	0	1
Prostate	4	6	13	10
Testis	0	1	2	0
Other Male Genital	1	0	1	1
Kidney	4	2	3	4
Bladder	3	2	2	3
Lymphopoietic	6	11	16	16
Brain	5	4	3	0
Connective Tissue	1	1	0	0
Malignant Melanoma	2	1	2	2
Other Skin	1	0	0	0
Other Endocrine	0	1	0	0
Bone	4	0	3	0
All Other Ill-Defined/Unspec.	7	5	4	8
All Tumor Sites	98	103	138	142

was for prostatic cancer, which had an odds ratio that was appreciably higher in the 15-24 year period than in the other periods.

Since most sites were slightly higher in the 15-24 or in the 20+ year period, these two periods were chosen for all subsequent analyses. The 20+ year latent period was chosen for cancers of the buccal cavity, esophagus, liver, lung, skin (melanoma),

TABLE 4

All Lung, Bronchus, and Trachea Cancer Deaths, 1957-1979: Adjusted Odds Ratios* According to Cigarette Smoking Habits Ascertained From Living Co-Workers

Cigarette Smoking Status (Packs/Day)	No. Cases	No. Controls	Crude Odds Ratio	Adjusted Odds Ratio	Lower 95% Confidence Limit	Upper 95% Confidence Limit	Two-Tailed P-Value
0	22	66	1.00	1.00	----	----	----
> 0-<1/2 or Exsmoker	21	30	2.10	2.19	0.94	5.10	0.07
1/2 - 1	62	45	4.13	5.01	2.33	10.74	0.01
1 - 2	26	16	4.88	4.38	1.85	10.35	0.01
> 2	27	10	8.10	10.34	3.66	29.19	0.01
Current, Unk. Am't.	5	2	7.50	5.55	0.93	33.26	0.06
Unknown	18	12	4.50	7.01	1.99	24.66	0.01
TOTAL	181	181					

* Odds ratios were adjusted for the matched-pairs design. The adjustment was carried out by the conditional logistic regression method for matched pairs (Breslow and Day, 1981).

TABLE 5

Adjusted* Relative Odds Ratios by Tumor Site and Exposure Period Prior To the Case's Death for Cases Occurring During the Interval 1957-1979

Cause of Death by Tumor Site++	No. Cases	Exposure Period (In Years Prior to The Case's Death)** 0-4	5-14	15-24	25+	15+	20+
Buccal cavity, Pharynx	7	1.00	1.00	1.00	1.00	1.00	1.00
Esophagus	9	0.00	1.00	0.00	0.50	1.00	1.00
Stomach	23	0.67	1.00	1.00	1.00	1.00	1.00
Colon and Rectum	54	0.44	0.54	0.75	0.50	0.69	0.45
Liver, Gall Bladder, Biliary Passages	10	----	>1	0.50	1.00	0.50	0.50
Pancreas	18	0.50	0.50	1.00	0.00	0.00	0.00
Larynx	8	0.00	1.00	1.00	0.00	2.00	0.00
Lung, Bronchus, Trachea	181	1.28	0.89	1.14	1.28	1.29	1.46
Prostate	33	0.00	3.00	6.00	1.67	2.50	2.50
Kidney	13	1.00	1.00	2.00	>1	2.00	1.00
Bladder	10	>1	>1	>1	>1	>1	>1
Lymphopoietic	49	0.00	0.67	0.67	0.50	0.50	0.60
Brain	12	0.25	0.00	0.50	0.00	0.50	0.00
Malignant Melanoma	7	----	1.00	1.00	----	0.00	1.00
Bone	7	2.00	2.00	2.00	1.00	2.00	1.00
Other Ill-Defined, Unspecified	24	1.00	1.00	1.67	0.33	0.83	0.67
All Sites	481	0.78	0.86	1.03	0.84	0.95	0.85

* Odds ratios were adjusted for the matched pairs design. Lung, bronchus and trachea cancer were also adjusted for cigarette smoking habits.

**In this analysis only exposure that occurred in the specified period prior to death of the case is considered to be relevant.

++In this analysis only tumor sites with more than three cases were included.

stomach, and bladder. The 15-24 year latent period was chosen for cancers of the pancreas, larynx, prostate, lymphopoietic system, brain, bone, colon and rectum, kidney, and all other ill-defined/unspecified sites.

Odds ratios were computed only for those sites with 7 or more cases. The one tumor that fell into the nose/nasal cavity/eustachian tube/middle ear category occurred in a worker who had no history of exposure to formaldehyde. As mentioned above, this worker's tumor was a eustachian tube/middle ear tumor rather than a nasal cancer.

Vital Status of the Controls

By definition, all of the study's cases were deceased. However, of the 481 controls, 270 were living and 211 were deceased at the time that work histories were abstracted.

In theory, the accuracy and completeness of work histories might vary according to the vital status of the employee. A lack of comparability between cases and controls with respect to data quality could introduce bias into our relative risk estimates.

To assess the actual influence of the control's vital status on study results, the data were split and analyzed separately according to the control's vital status. For an analysis of ever-versus-never exposed to formaldehyde, the adjusted (matched) odds ratio for all cancer deaths was 0.80 for living controls and 0.90 for deceased controls (Table 6). This analysis suggests that the control's vital status had no appreciable impact on the study's results.

TABLE 6

Odds Ratios for All Cancer Deaths According to Whether Employee Ever Worked around Formaldehyde and According to the Vital Status of the Control at the Time the Work Histories were Abstracted: Latent Period = 20+ Years

Matched Pairs in Which the Control Was Living				
Exposure	No. Cases	No. Controls	Crude Odds Ratio	Adjusted Odds Ratio
Background	214	207	1.00	1.00
Ever Exposed	56	63	0.86	0.80

Matched Pairs in Which the Control Was Deceased				
Exposure	No. Cases	No. Controls	Crude Odds Ratio	Adjusted Odds Ratio
Background	173	170	1.00	1.00
Ever Exposed	38	41	0.91	0.90

Source of Work History

To assess the impact that source of work history had on study results, the data were analyzed according to whether the work history was based on records or on memory (Table 7).

In about 5% (26) of the matched pairs, both case and control work histories were based on memory. In about 80% (389) of the matched pairs, both case and control work histories were based on records. There were no consistent differences in the adjusted (matched) odds ratios according to source of work history. The adjusted odds ratio for pairs that used records for both case and control was 0.90, compared to an adjusted odds ratio of 0.80 for

TABLE 7

Odds Ratios for All Cancer Deaths According to Whether Employee Ever Worked around Formaldehyde and According to Source of Work History: Latent Period = 20+ Years

Source of Work History:
Case = Records; Control = Records

Exposure	No. Cases	No. Controls	Crude Odds Ratio	Adjusted Odds Ratio
Background	315	311	1.00	1.00
Ever Exposed	74	78	0.94	0.92

Source of Work History
Case = Records; Control = Memory

Exposure	No. Cases	No. Controls	Crude Odds Ratio	Adjusted Odds Ratio
Background	30	29	1.00	1.00
Ever Exposed	2	3	0.64	0.67

Source of Work History
Case = Memory; Control = Memory

Exposure	No. Cases	No. Controls	Crude Odds Ratio	Adjusted Odds Ratio
Background	14	16	1.00	1.00
Ever Exposed	12	10	1.37	1.50

Source of Work History
Case = Memory; Control = Records

Exposure	No. Cases	No. Controls	Crude Odds Ratio	Adjusted Odds Ratio
Background	28	21	1.00	1.00
Ever Exposed	6	13	0.35	0.30

pairs where memory was used for one or both subjects' work histories. This analysis indicates that source of work history had no appreciable impact on the study's results.

Payclass

To determine whether there was interaction between formaldehyde exposure and payclass, wage roll was analyzed separately from salary roll. The odds ratios for ever-versus-never exposed to formaldehyde differed little between wage and salary workers. For the analysis of all cancer deaths (Table 8), the matched-pairs odds ratios were 0.88 and 0.73 for wage and salary roll workers, respectively. Thus, there was no indication of a payclass-exposure interaction.

Duration of Exposure

To see if cancer risk varied with duration of exposure, subjects with any exposure potential were classified as to whether they had worked around formaldehyde for less than 5 years, or for 5 years or more. Exposure duration was calculated after allowing for a latent period between exposure and disease (see METHODS section).

The data showed no significant relationships between exposure duration and the relative odds ratios (Tables 9 and 10). None of the odds ratios was significantly higher than 1.0. However, the cancer odds ratio for prostate (Table 9, OR = 4.8) and bladder (Table 10, OR = 7.0) were slightly elevated in the 5 or more year duration category.

TABLE 8

Distribution of Matched Pairs for All Cancer Deaths According to Payclass: Ever Vs. Never Exposed -- Latent Period = 15-24 Years

Wage	Cases: Not Exposed	Cases: Exposed	Comment
Controls: Not Exposed	242	45	Matched-Pairs Odds Ratio = 0.88 % Exposed = 21%
Controls: Exposed	51	29	

SALARY	Cases: Not Exposed	Cases: Exposed	Comment
Controls: Not Exposed	79	11	Matched-Pairs Odds Ratio = 0.73 % Exposed = 19%
Controls: Exposed	15	9	

W&S	Cases: Not Exposed	Cases: Exposed	Comment
Controls: Not Exposed	321	56	Matched-Pairs Odds Ratio = 0.85 % Exposed = 21%
Controls: Exposed	66	38	

TABLE 9

Cancer Deaths at Study Plants, 1957-1979*
Crude Odds Ratios** According to Number of Years Employee
Worked Around Formaldehyde: Latent Period = 15-24 Years

Organ	No. Cases	No. Controls	Odds Ratio
Pancreas			
None	16	16	1.00
Less than 5	0	0	----
5 or More	2	2	1.00
Larynx			
None	6	6	1.00
Less than 5	2	1	2.00
5 or More	0	1	0.00
Prostate			
None	25	30	1.00
Less than 5	0	1	0.00
5 or More	8	2	4.80
Lymphopoietic System			
None	43	41	1.00
Less than 5	2	6	0.32
5 or More	4	2	1.91
Brain and Other Nervous System			
None	11	10	1.00
Less than 5	1	0	----
5 or More	0	2	0.00

Organ	No. Cases	No. Control	Odds Rati
Bone			
None	4	5	1.0
Less than 5	1	0	----
5 or More	2	2	1.2
All Other Ill-Defined/Unspecified			
None	18	20	1.0
Less than 5	1	0	----
5 or More	5	4	1.3
Colon and Rectum			
None	42	39	1.0
Less than 5	4	3	1.2
5 or More	8	12	0.6
Kidney			
None	11	12	1.0
Less than 5	1	0	---
5 or More	1	1	1.0
All Sites			
None	381	383	1.0
Less than 5	33	29	1.1
5 or More	67	69	0.9

* Analyses are restricted to sites with 7 or more cancer deaths.
**Odds ratios estimate risk of less than 5 years exposure and 5 or more year exposure relative to the risk of no exposure.

TABLE 10

Cancer Deaths at Study Plants, 1957-1979*
Crude Odds Ratios** According to Number of Years Employee
Worked Around Formaldehyde: Latent Period = 20+ Year

Organ	No. Cases	No. Controls	Odds Ratio	Organ	No. Cases	No. Controls	Odds Ratio
Buccal Cavity				Malignant Melanoma			
None	6	6	1.00	None	6	6	1.00
Less than 5	0	0	----	Less than 5	1	1	1.00
5 or More	1	1	1.00	5 or More	0	0	----
Esophagus				Stomach			
None	6	6	1.00	None	17	17	1.00
Less than 5	2	1	2.00	Less than 5	0	0	----
5 or More	1	2	0.50	5 or More	6	6	1.00
Liver, Gall Bladder, Bile Duct				Bladder			
None	9	8	1.00	None	4	7	1.00
Less than 5	0	1	0.00	Less than 5	2	2	1.75
5 or More	1	1	0.89	5 or More	4	1	7.00
Lung, Bronchus, Trachea				All Sites			
None	142	142	1.00	None	387	377	1.00
Less than 5	24	20	1.20	Less than 5	39	46	0.83
5 or More	15	19	0.79	5 or More	55	58	0.92

* Analyses are restricted to sites with 7 or more cancer deaths.

**Odds ratios estimate risk of less than 5 years exposure and 5 or more years exposure relative to the risk of no exposure.

Frequency of Exposure: Continuous vs Intermittent

The risks of continuous and intermittent exposures relative to no exposure were assessed in Tables 11 and 12. For no cancer site was the odds ratio significantly higher than 1.0 for either continuous or intermittent exposure categories.

Frequency and Level of Exposure

The potential for formaldehyde exposure was further defined according to levels of exposure within the intermittent and continuous categories. The data showed no pattern of increasing risk with increasing dose (Tables 13 and 14). Because of the large number of exposure categories, these and subsequent analyses were restricted to tumor sites with 5 or more exposed cancer deaths. Continuous Level 3 was combined with Continuous Level 2 since there were only 2 cases and 2 controls in the Level 3 category.

Cumulative Exposure Index

As a means of simultaneously examining exposure duration and exposure level, a cumulative exposure index (CEI) was computed for each case and control. CEI was computed as the product of the exposure duration and exposure level. Duration of exposure was computed after allowing for a latent period (see METHODS). Level of exposure was an exposure rank assigned to a given exposure category.

Although a variety of plausible exposure ranking schemes were tried, it made little difference which ranking scheme was used to compute the CEI. The results were very similar regardless of whether intermittent or continuous-direct exposures were assumed to be more significant biologically. The exposure ranking scheme that was used in Tables 15 and 16 and which showed the strongest dose-response trend was as follows:

Exposure Category	Rank
Background	0
Intermittent #1	2
Intermittent #2	4
Continuous #1	1
Continuous #2	3
Continuous #3	5

The data showed no significant relationships between CEI and cancer risk. As in the analyses of exposure duration, the odds ratios for prostatic (OR=3.6) and bladder (OR=5.2) cancers were slightly but not significantly higher than 1.0 for the highest CEI category (Tables 15 and 16).

Year of Death

Sometimes cancer risks will show trends with time that reflect dates of introduction or changing usage patterns of a physical agent. To investigate this possibility, the data were split into 5-year calendar periods.

No significant trends with time were observed (Table 17). For this and subsequent analyses, only two sites were presented: lung and prostate. Lung was included because it was a site of special interest given the results of the CIIT study in rats. Prostate was presented because this was the site that in this

TABLE 11

Cancer Deaths at Study Plants, 1957-1979*
Crude Odds Ratios** According to Whether Employee Had Continuous or Intermittent Formaldehyde Exposure: Latent Period = 15-24 Years

Organ	No. Cases	No. Controls	Odds Ratio	Organ	No. Cases	No. Controls	Odds Ratio
Pancreatic				Lymphopoietic			
None (Background)	16	16	1.00	None (Background)	43	41	1.00
Continuous Only	1	1	1.00	Continuous Only	3	3	0.95
Intermittent Only	1	1	1.00	Intermittent Only	3	5	0.57
Larynx				All Other Ill-Defined/Unspecified			
None (Background)	6	6	1.00	None (Background)	11	10	1.00
Continuous Only	2	1	2.00	Continuous Only	0	1	0.00
Intermittent Only	0	1	0.00	Intermittent Only	1	1	0.91

Prostatic			
None (Background)	25	30	1.00
Continuous Only	2	1	2.40
Intermittent Only	6	2	3.60
Kidney			
None (Background)	11	12	1.00
Continuous Only	0	0	----
Intermittent Only	2	1	2.18
Colon and Rectum			
None (Background)	42	39	1.00
Continuous Only	4	5	0.74
Intermittent Only	8	10	0.74
Bone			
None (Background)	4	5	1.00
Continuous Only	1	0	----
Intermittent Only	2	2	1.25
All Other Ill-Defined/Unspecified			
None (Background)	18	20	1.00
Continuous Only	3	2	1.67
Intermittent Only	3	2	1.67
All Sites			
None (Background)	381	383	1.00
Continuous Only	36	35	1.03
Intermittent Only	62	61	1.02
Cont. & Intermit.	2	2	1.01

* Analyses are restricted to sites with 7 or more cancer deaths.

**Odds ratios estimate risk of less continuous exposure and of intermittent exposure relative to the risk of no exposure.

TABLE 12

Cancer Deaths at Study Plants, 1957-1979*
Crude Odds Ratios** According to Whether Employee Had Continuous or Intermittent Formaldehyde Exposure: Latent Period = 20 or More Years

Organ	No. Cases	No. Controls	Odds Ratio	Organ	No. Cases	No. Controls	Odds Ratio
Buccal Cavity				Bladder			
None (Background)	6	6	1.00	None (Background)	4	7	1.00
Continuous Only	0	0	----	Continuous Only	3	1	5.25
Intermittent Only	1	1	----	Intermittent Only	3	2	2.63
Esophagus				Malignant Melanoma			
None (Background)	6	6	1.00	None (Background)	6	6	1.00
Continuous Only	2	1	1.00	Continuous Only	1	0	----
Intermittent Only	1	1	1.00	Intermittent Only	0	0	----
				Cont. & Intermit.	0	1	0.00

Stomach			
None (Background)	17	17	1.00
Continuous Only	2	2	1.00
Intermittent Only	3	4	1.00
Cont. & Intermit.	1	0	----
Liver, Gall Bladder, Bile Duct			
None (Background)	9	8	1.00
Continuous Only	0	0	----
Intermittent Only	1	2	0.44

Lung, Bronchus, Trachea			
None (Background)	142	142	1.00
Continuous Only	17	14	1.21
Intermittent Only	22	24	0.92
Cont. & Intermit.	0	1	0.00
All Sites			
None (Background)	387	377	1.00
Continuous Only	39	32	1.19
Intermittent Only	54	70	0.75
Cont. & Intermit.	1	2	0.49

* Analyses are restricted to sites with 7 or more cancer deaths.

**Odds ratios estimate risk of less continuous exposure and of intermittent exposure relative to the risk of no exposure.

TABLE 13

Cancer Deaths at Study Plants, 1957-1979*:
Crude Odds Ratios According to Whether Employee Had Intermittent (Levels 1-2) or Continuous (Levels 1-3) Formaldehyde Exposure: Latent Period = 15-24 Years

Organ	No. Cases	No. Controls	Odds Ratio	Organ	No. Cases	No. Controls	Odds Ratio
Colon & Rectum				Prostate			
None (Background)	42	39	1.00	None (Background)	25	30	1.00
Intermittent #1	3	5	0.56	Intermittent #1	4	2	2.40
Intermittent #2	5	4	1.16	Intermittent #2	2	0	----
Continuous #1	2	1	1.86	Continuous #1	1	0	----
Continuous #2/#3	1	3	0.31	Continuous #2	1	1	----
Mixed	1	2	0.46	Mixed	0	0	----

Lymphopoietic System			
None (Background)	43	41	1.00
Intermittent #1	2	3	0.64
Intermittent #2	1	2	0.48
Continuous #1	2	0	----
Continuous #2/#3	1	3	0.32
Mixed	0	0	----

Other Ill-Defined/Unspecified Sites			
None (Background)	18	20	1.00
Intermittent #1	0	1	0.00
Intermittent #2	3	1	3.33
Continuous #1	1	0	----
Continuous #2/#3	1	2	0.56
Mixed	1	0	----

All Sites			
None (Background)	381	383	1.00
Intermittent #1	35	36	0.98
Intermittent #2	27	23	1.18
Continuous #1	10	7	1.44
Continuous #2	18	23	0.79
Continuous #3	2	2	1.01
Mixed	8	7	1.15

Intermittent #1: 2.0 ppm.
Intermittent #2: 2.0 ppm.
Continuous #1: 0.1 ppm.
Continuous #2: 0.1 2.0 ppm.
Continuous #3 2.0 ppm.
Mixed: Exposure occurred in more than one exposure category.

---

* Analyses are restricted to sites with 5 or more exposed cancer deaths.

TABLE 14

Cancer Deaths at Study Plants, 1957-1979*:
Crude Odds Ratios According to Whether Employee Had Intermittent (Levels 1-2)
or Continuous (Levels 1-3) Formaldehyde Exposure: Latent Period = 20 or More Years

Organ	No. Cases	No. Controls	Odds Ratio	Organ	No. Cases	No. Controls	Odds Ratio
Lung				Stomach			
None (Background)	142	142	1.00	None (Background)	17	17	1.00
Intermittent #1	14	16	0.88	Intermittent #1	3	2	1.50
Intermittent #2	8	5	1.60	Intermittent #2	0	2	0.00
Continuous #1	5	3	1.67	Continuous #1	0	1	0.00
Continuous #2/#3	11	10	1.10	Continuous #2/#3	1	1	1.00
Mixed	1	5	0.20	Mixed	2	0	----

Bladder			
None (Background)	4	7	1.00
Intermittent #1	1	1	1.75
Intermittent #2	1	0	----
Continuous #1	0	0	----
Continuous #2/#3	2	1	3.50
Mixed	2	1	3.50

All Sites			
None (Background)	387	377	1.00
Intermittent #1	30	43	0.68
Intermittent #2	20	19	1.03
Continuous #1	11	8	1.34
Continuous #2/#3	21	23	0.89
Mixed	12	11	1.06

Intermittent #1: $<$ 2.0 ppm.
Intermittent #2: $\geq$ 2.0 ppm.
Continuous #1: $<$ 0.1 ppm.
Continuous #2: 0.1 – 2.0 ppm.
Continuous #3 $\geq$ 2.0 ppm.
Mixed: Exposure occurred in more than one exposure category.

* Analyses are restricted to sites with 5 or more <u>exposed</u> cancer deaths.

TABLE 15

Cancer Deaths at Study Plants, 1957-1979*:
Crude Odds Ratios For Formaldehyde Exposure as Measured by the Cumulative Exposure Index (CEI)**: Latent Period = 15-24 Years

Organ	No. Cases	No. Controls	Odds Ratio	Organ	No. Cases	No. Controls	Odds Ratio
Colon and Rectum				Prostate			
0 (Background only)	42	39	1.00	0 (Background only)	25	30	1.00
≤20	6	8	0.70	≤20	5	2	3.00
>20	6	7	0.80	>20	3	1	3.60
Lymphopoietic Stystem				All Sites			
0 (Background only)	43	41	1.00	0 (Background only)	381	383	1.00
≤20	5	7	0.68	≤20	65	62	1.05
>20	1	1	0.95	>20	35	36	0.98

* Analyses are restricted to sites with 5 or more exposed cancer deaths.
**Cumulative exposure (CEI) = (no. years in a given exposure category) x (rank of that category).
Formaldehyde exposure potentials were assigned to the following ranks before the CEI was computed:

Potential:	Background	Interm. #1	Interm. #2	Contin. #1	Contin. #2	Contin. #3
Rank:	0	2	4	1	3	5

TABLE 16

Cancer Deaths at Study Plants, 1957-1979*:
Crude Odds Ratios For Formaldehyde Exposure as Measured by the Cumulative Exposure Index (CEI)**: Latent Period = 20 or More Years

Organ	No. Cases	No. Controls	Odds Ratio	Organ	No. Cases	No. Controls	Odds Ratio
Lung, Bronchus				Stomach			
0 (Background only)	142	142	1.00	0 (Background only)	17	17	1.00
≤ 20	27	25	1.08	≤ 20	1	1	1.00
> 20	12	14	0.86	> 20	5	5	1.00
Bladder				All Sites			
0 (Background only)	4	7	1.00	0 (Background only)	387	377	1.00
≤ 20	3	2	2.63	≤ 20	48	58	0.81
> 20	3	1	5.25	> 20	46	46	0.97

* Analyses are restricted to sites with 5 or more exposed cancer deaths.

**Cumulative exposure (CEI) = (no. years in a given exposure category) x (rank of that category).
Formaldehyde exposure potentials were assigned to the following ranks before the CEI was computed:

Potential:	Background	Interm. #1	Interm. #2	Contin. #1	Contin. #2	Contin. #3
Rank:	0	2	4	1	3	5

TABLE 17

Cancer Deaths at Study Plants, 1957-1979:
Crude Odds Ratios According to Whether Employee Ever Worked Around
Formaldehyde--Data Broken Down by Year of Death: Latent Period = 20+ Years
For Lung Cancer; 15-24 Years for Prostatic Cancer and All Cancer

	1957-64			1965-69			1970-74		
Organ	No. Cases	No. Controls	Odds Ratio	No. Cases	No. Controls	Odds Ratio	No. Cases	No. Controls	Odds Ratio
Lung, Bronchus									
Background	23	23	1.00	37	35	1.00	41	40	1.00
Ever Exposed	4	4	1.00	9	11	0.77	9	10	0.88
Prostate									
Background	4	4	1.00	5	6	1.00	10	12	1.00
Ever Exposed	0	0	----	1	0	----	3	1	3.60
All Sites									
Background	79	73	1.00	83	81	1.00	114	114	1.00
Ever Exposed	19	25	0.70	20	22	0.89	24	24	1.00

Organ	1975–79			1957–79		
	No. Cases	No. Controls	Odds Ratio	No. Cases	No. Controls	Odds Ratio
Lung, Bronchus						
Background	41	44	1.00	142	142	1.00
Ever Exposed	17	14	1.30	39	39	1.00
Prostate						
Background	6	8	1.00	25	30	1.00
Ever Exposed	4	2	2.67	8	3	3.20
All Sites						
Background	105	115	1.00	381	383	1.00
Ever Exposed	37	27	1.50	100	98	1.03

study has shown the strongest although statistically nonsignificant relationship to formaldehyde exposure.

Age at Death

Sometimes a carcinogen may accelerate spontaneously occurring biologic processes so as to produce cancer at younger ages than usual but to cause little or no increase in overall cancer rates. To determine if risks were unusually high in younger age groups, the data were split into two groups according to the case's age at death.

The odds ratios in younger age groups were not significantly different from the odds ratios in older age groups (Table 18). There was no tendency for odds ratios to be higher in younger as opposed to older age groups.

Year of First Exposure

An analysis by year of first exposure will sometimes reveal unusually high risks for workers who were first exposed during a certain time interval. Such excesses may reflect dates of introduction or changes in exposure patterns for a physical agent.

When the formaldehyde data were analyzed by year of first exposure, no increasing or decreasing trends in time were noted for lung or prostate cancer risk (Table 19).

Age at First Exposure

Certain investigators have suggested that humans may be more susceptible to the effects of carcinogens at earlier ages than at

TABLE 18

Cancer Deaths at Study Plants, 1957-1979:
Crude Odds Ratios According to Whether Employee Ever Worked Around Formaldehyde--Data Broken Down by Age of Death: Latent Period = 20+ Years For Lung Cancer, 15-24 Years for Prostatic Cancer and All Cancer

	Ages 20-64 Years			Ages 65 Years or Older		
Organ	No. Cases	No. Controls	Odds Ratio	No. Cases	No. Controls	Odds Ratio
Lung, Bronchus						
Background	83	80	1.00	59	62	1.00
Ever Exposed	22	25	0.85	17	14	1.28
Prostate						
Background	2	2	1.00	23	28	1.00
Ever Exposed	1	1	1.00	7	2	4.26
All Sites						
Background	185	180	1.00	196	203	1.00
Ever Exposed	50	55	0.88	50	43	1.20

later ages. To assess this possibility the data have been analyzed according to age at first exposure.

There was no suggestion that relative risk was higher among workers first exposed at earlier ages than among those first exposed at later ages (Table 19). In fact, controls tended to be first exposed at earlier ages than were the cases.

Interaction With Smoking Habits

There have been a number of reports indicating that tobacco use can act in combination with other hazardous agents in the work place to produce or increase the severity of a wide variety of adverse health effects. The asbestos-smoking-lung cancer relationship is the most publicized example of this interaction.

To investigate the possibility of interaction between smoking, formaldehyde, and lung cancer, an interaction (cross product) term was introduced into the logistic analysis. The results of this analysis showed no evidence for an interaction between smoking and formaldehyde.

Furthermore, when the lung cancer data were stratified according to smoking habits, there was little difference in the formaldehyde exposure odds ratio from one smoking category to another (Table 20).

Plant Site-Specific Analyses

Plant site-specific analyses (Tables 21 and 22) showed that no plant's overall cancer or lung cancer odds ratio for formaldehyde exposure was significantly greater than 1.0. That is, for any given plant the cancer risk in formaldehyde workers

TABLE 19

Cancer Deaths 1957-1979: Odds Ratios
According to Year and Age of First Formaldehyde Exposure

Organ and Year First Exposed	No. Cases	No. Controls	Crude Odds Ratio	Organ and Age First Exposed	No. Cases	No. Controls	Crude Odds Ratio
Lung, Bronchus				Lung, Bronchus			
Never Exposed	119	122	1.00	Never Exposed	119	122	1.00
Prior to 1940	7	3	2.39	Less than 30 Years	8	9	0.91
1940-1949	17	26	0.67	30-39 Years	13	16	0.83
1950-1959	23	12	1.96	40-49 Years	23	18	1.31
1960-1969	15	18	0.85	50 or Over	18	16	1.15
Prostate				Prostate			
Never Exposed	24	27	1.00	Never Exposed	24	27	1.00
Prior to 1940	2	1	2.25	Less than 30 Years	1	2	0.56
1940-1949	4	3	1.50	30-39 Years	1	1	1.13
1950-1959	3	1	3.38	40-49 Years	3	1	3.38
1960-1969	0	1	0.00	50 or Over	4	2	2.25
All Sites				All Sites			
Never Exposed	339	328	1.00	Never Exposed	339	328	1.00
Prior to 1940	22	19	1.12	Less than 30 Years	19	26	0.71
1940-1949	40	50	0.77	30-39 Years	32	38	0.81
1950-1959	58	49	1.15	40-49 Years	48	43	1.08
1960-1969	22	35	0.61	50 or Over	43	46	0.90

TABLE 20

Lung, Bronchus, and Trachea Cancer Deaths, All Ages, 1959-1979: Crude Odds Ratios According to Whether Employee Ever Worked Around Formaldehyde and According to Cigarette Smoking Status: Latent Period = 20+ Years

Formaldehyde Exposure Potential	Nonsmokers, Ex-smokers, and Light Smokers* No. Cases	No. Controls	Odds Ratio
Background Only	35	71	1.0
Ever Exposed	8	25	0.7
Total	43	96	

Formaldehyde Exposure Potential	Moderate and Heavy Cigarette Smokers** No. Cases	No. Controls	Odds Ratio
Background Only	91	59	1.0
Ever Exposed	24	12	1.3
Total	115	71	

* Smoked less than 1/2 pack per day.
**Smoked at least 1/2 pack per day.

TABLE 21

All Cancer Deaths at Study Plants, 1957-1979--Crude Ratios by Plant According to Ever *vs*. Never Exposed to Formaldehyde: Latent Period = 20+ Years

Plant Site	No. Cases	No. Controls	Odds Ratio
Parlin F&F			
Background	59	60	1.0
Exposed	20	19	1.1
Belle			
Background	196	188	1.0
Exposed	22	30	0.7
Parlin Photo			
Background	64	58	1.0
Exposed	19	25	0.7
Toledo F&F			
Background	8	9	1.0
Exposed	11	10	1.2
Toledo C&P			
Background	3	1	1.0
Exposed	0	2	<1.0
Perth Amboy			
Background	3	3	1.0
Exposed	7	7	1.0
Rochester Photo			
Background	21	24	1.0
Exposed	14	11	1.5
Washington Works			
Background	33	34	1.0
Exposed	1	0	---

TABLE 22

Lung Cancer Deaths at Study Plants, 1957-1979--Crude Odds Ratios by Plant According to Ever *vs*. Never Exposed to Formaldehyde: Latent Period = 20+ Years

Plant Site	No. Cases	No. Controls	Odds Ratio
Parlin F&F			
Background	14	13	1.0
Exposed	5	6	0.8
Belle			
Background	87	89	1.0
Exposed	17	15	1.2
Parlin Photo			
Background	14	14	1.0
Exposed	6	6	1.0
Toledo F&F			
Background	4	4	1.0
Exposed	6	6	1.0
Toledo C&P			
Background	2	0	1.0
Exposed	0	2	<1.0
Perth Amboy			
Background	2	3	1.0
Exposed	2	1	3.0
Rochester Photo			
Background	7	7	1.0
Exposed	3	3	1.0
Washington Works			
Background	12	12	1.0
Exposed	0	0	---

was not significantly higher than the risk in nonexposed co-workers.

DISCUSSION

The significance of these results is better appreciated after a discussion of the study's strengths and potential limitations.

Strengths

a. Sensitivity

In a statistical sense, the study was very sensitive to increases in lung cancer and overall cancer risks among exposed workers. In fact, the study's sensitivity was comparable to what could be expected from a cohort study of 2650 formaldehyde workers followed for at least 20 years.

- The study had an 80% chance of detecting at least a 1.5-fold increase in overall cancer if formaldehyde indeed caused such an excess (Table 23).
- The study had an 80% chance of detecting at least a 2.3-fold excess of lung cancer among workers with 5 or more years of exposure if formaldehyde had indeed caused such an excess (Table 24).

b. Internal Comparisons

The case-control study by design compared cancer risk in formaldehyde-exposed workers to cancer risk in nonexposed co-workers. Furthermore, the study compared only workers who were of the same birth cohort, geographic area, plant site, payclass, and adjusted service date. Thus, this study was not

TABLE 23

Sensitivity of the Formaldehyde Study by Tumor Site: The Smallest Detectable Relative Odds Ratios for Workers Ever Exposed to Formaldehyde: Latent Period = 20+ Years

Cause of Death by Tumor Site	No. Cases	No. Controls	Smallest Detectable Odds Ratio	
			80% Power*	90% Power*
Buccal Cavity, Pharynx	7	7	30.7	65.8
Esophagus	9	9	18.0	29.5
Stomach	23	23	5.9	7.5
Colon and Rectum	54	54	3.3	3.9
Liver, Gall Bladder, Biliary Passages	10	10	15.1	23.4
Pancreas	18	18	7.3	9.7
Larynx	8	8	22.5	40.3
Lung, Bronchus, Trachea	181	181	2.0	2.2

Prostate	33	33	4.4	5.5
Kidney	13	13	10.5	14.8
Bladder	10	10	15.1	23.4
Lymphopoietic	49	49	3.5	4.1
Brain	12	12	11.6	16.8
Malignant Melanoma	7	7	30.7	65.8
Bone	7	7	30.7	65.8
Other Ill-Defined, Unspecified	24	24	5.6	7.2
All Sites	481	481	1.5	1.6

* If the population odds ratio is at least as large as that given, the probability of declaring the sample odds ratio to be statistically significant is 0.80 for 80% power, 0.90 for 90% power. Statistical significance tests are assumed to be two-tailed, and significance is judged at the 0.05 probability level. Twenty percent of the controls are assumed to have been exposed to formaldehyde.

TABLE 24

Sensitivity of the Formaldehyde Study by Tumor Site: The Smallest Detectable Relative Odds Ratios for Workers Exposed For at Least Five Years: Latent Period = 20+ Years

Cause of Death by Tumor Site	No. Cases	No. Controls	Smallest Detectable Odds Ratio for 80% Power*
Buccal Cavity, Pharynx	7	7	32.4
Esophagus	9	9	21.6
Stomach	23	23	7.7
Colon and Rectum	54	54	4.2
Liver, Gall Bladder, Biliary Passages	10	10	18.7
Pancreas	18	18	9.7
Larynx	8	8	25.9
Lung, Bronchus, Trachea	181	181	2.3

Prostate	33	33	5.8
Kidney	13	13	13.6
Bladder	10	10	18.7
Lymphopoietic	49	49	4.4
Brain	12	12	14.9
Malignant Melanoma	7	7	32.4
Bone	7	7	32.4
Other Ill-Defined, Unspecified	24	24	7.5
All Sites	481	481	1.7

* If the population odds ratio is at least as large as that given, the probability of declaring the sample odds ratio to be statistically significant is 0.80 for 80% power. Statistical significance tests are assumed to be two-tailed, and significance is judged at the 0.05 probability level. Ten percent of the controls are assumed to have been exposed to formaldehyde for at least 5 years.

TABLE 25

The Effect of Exposure Misclassification Errors* on the Observed Odds Ratio**

Error Rates				True Odds Ratio	Observed Odds Ratio
False Pos. in Cases	False Pos. in Controls	False Neg. in Cases	False Neg. in Controls		
0.10	0.10	0	0	3.0	2.4
0	0	0.10	0.10	3.0	2.9
0.05	0.05	0.10	0.10	3.0	2.5
0	0	0.05	0.10	3.0	3.1
0.10	0.05	0	0	3.0	3.0
0	0	0.10	0.05	3.0	2.7
0.05	0.10	0	0	3.0	2.2
0.10	0.05	0.05	0.10	3.0	3.1
0.10	0.05	0.10	0.05	3.0	2.7
0.05	0.10	0.05	0.10	3.0	2.2
0.05	0.10	0.10	0.05	3.0	1.9

* Copeland et al.[9] (1977).

**Assumptions: 1) True odds ratio = 3.0; 2) 20% of the controls were exposed to formaldehyde; 3) the false positive and false negative error rats were on the order of 5-10%.

subject to the uncertainties characteristic of many other studies concerning the importance of the healthy worker effect or of geographic, socioeconomic, or temporal differences in cancer rates.

c. Smoking Habits

Cigarette smoking habits were ascertained by co-worker proxy for cases and control. This feature allowed the potential confounding effects of cigarette smoking habits to be reduced so that the relative risks of formaldehyde alone could be assessed. This adjustment for smoking habits greatly strengthened our conclusions concerning the relationship between lung cancer and formaldehyde exposure.

The validity of the smoking histories was supported by the strong dose-response relationship found for lung cancer and cigarette smoking (Table 4). This relationship was similar to what has been found nationally in cohort studies of smoking and lung cancer.

d. Exposure Frequency, Intensity, and Duration

The data were such that meaningful comparisons could be made for frequency, intensity, and duration of exposure. Forty percent of exposed workers fell into the continuous exposure category and 60% fell into the intermittent exposure category. Two-thirds of the continuous exposure category had been exposed at 0.1 ppm or above, and 45% of the intermittent exposure category had been exposed at 2 ppm or above.

e. Adequate Latent Period

First exposures to formaldehyde occurred far enough in the past to allow for an adequate latent period. About 45% of the exposed cases and controls were first exposed prior to the 1940's when formaldehyde levels in many instances were higher than they are today. All of these workers had at least a 17-year latent period from the time of first exposure until death from cancer. Many had longer latent periods. Thus, the negative findings of this study cannot be said to be due to too short a latent period.

Limitations

a. Low Sensitivity for Extremely Rare Cancers

The study had limited sensitivity for cancer sites for which there were only a few cases. Even so, the observation that there were no formaldehyde-exposed nasal cancers in this study or in other studies permits us to establish an upper limit on the possible risk of nasal cancer from formaldehyde exposure.

b. Nonpensioned Terminations, Transfers

Cancer deaths among employees who left the Company without a pension or who transferred to a plant not in the study were not included in the study. Based on the experience of other cohort studies at Du Pont, these exclusions might comprise about 15-20% of the underlying work force. The exclusion of these subjects would bias the results only if the reasons for leaving the plant were related both to past exposures and to the risk of dying from cancer.

Even if these conditions were met, the extent of the bias was most likely minimal. It is easy to show that cancer risks in formaldehyde workers excluded from the study would have to be 5 to 10 times higher than the risks in those included before there would be a noticeable effect on the study's results. In short, it is unlikely that the exclusion of nonpensioned terminations and transfers could account for the study's negative findings.

c. Mixed Exposures

At all of the plants studied, workers were exposed to a variety of materials in addition to formaldehyde. These mixed exposures could be viewed as a strength or as a limitation of the study.

With respect to the issue of synergism between formaldehyde and other chemicals, the presence of mixed exposures was a strength. In this study the cancer risk among workers exposed both to formaldehyde and to other chemicals was no higher than the risk in co-workers with no formaldehyde exposure. This finding suggests that synergism between formaldehyde and other materials in the work place is not a problem.

On the other hand, mixed exposures could have been a limitation if cancer risks in workers not exposed to formaldehyde were unusually high due to exposures to other occupational carcinogens. If formaldehyde were a carcinogen, nonformaldehyde workers' exposures to other carcinogens could have masked the effects of formaldehyde.

Haskell Laboratory has performed risk assessments on several materials that are used at the study plants. These materials are listed in Appendix C.

d. Cancer Mortality vs Incidence

Since cancer deaths rather than incident cases were studied, the issue of survival rates becomes important. For cancers such as lung or pancreas, the survival rates are so poor that a study of mortality usually leads to the same conclusions as a study of incidence. For sites such as skin in which survival is very good, excesses of cancer incidence may not be accompanied by a corresponding increase in cancer mortality. The five-year survival rate for nasal cancer, which is about 45%, is intermediate between the survival rates for lung and skin cancer, so that an increase in nasal cancer incidence should be reflected by a subsequent increase in its mortality.

e. Qualitative Misclassification of Exposure

Qualitative misclassification of exposure could have biased the study's results and caused either an overestimation or underestimation of the odds ratio. The direction and extent of this bias depends on the nature and degree of the misclassification.

A quantitative analysis of the situation (Table 25) showed that even if misclassification were as high as 5-10%, it was very unlikely that the bias could have been extensive enough to account for finding no association between formaldehyde exposure and cancer[9].

f. Quantitative Misclassification of Exposure

Estimates of exposure levels were based in part on air monitoring data from current working conditions. However, our knowledge of past process changes, engineering controls, odor and sensory irritation thresholds, and anecdotal reports by long-service employees suggest that formaldehyde levels were higher in the past than they are today.

When exposure levels were assigned to the various jobs, there may have been a tendency to emphasize current conditions and de-emphasize the past. Such a tendency could have caused systematic underestimates of the levels to which formaldehyde workers were exposed.

g. Smoking Histories from Proxies

Smoking histories of cases and controls were primarily ascertained through co-worker proxies. The ideal source of smoking history is the worker himself. If proxies must be used, the subject's next of kin are generally preferred. Co-workers will usually be unfamiliar with a subject's off-the-job smoking habits and with some of the finer details of his smoking history.

Despite this limitation, the study's lung cancer data showed a strong dose-response relationship with the amount smoked per day. This finding suggests that the smoking data were good enough to remove most of the confounding effects of cigarette smoking.

SIGNIFICANCE

The CIIT study suggested that if formaldehyde were a human carcinogen, then the upper respiratory tract, particularly the nasal cavities, were the most likely targets in humans. Some critics have argued that the lower respiratory tract should also be considered a likely target, since rats are obligate nose breathers but man is not.

The present study and three other epidemiology studies (Wong,[5] Walrath,[3] and Marsh[4]) of workers exposed to formaldehyde have found no excess risks for cancers of the nasal cavities, respiratory tract, or buccal cavity that could be attributed to formaldehyde exposure.

Walrath[3] found slight excesses for skin, kidney, and brain cancers, these excesses were not confirmed by the present study. Little importance should be attached to the positive findings in the Walrath study, since it was proportionate mortality (PMR) and subject to the limitations common to all PMR studies. Furthermore, Walrath's reported excess of skin cancers was obtained only after several histologic types with differing etiology were combined.

Wong[5] found a slight excess of prostate cancer among exposed workers. Coincidentally, cancer of the prostate was one of two sites in the present study that showed slight although statistically nonsignificant elevations in the odds ratio. The bladder was the other site that showed a slight excess.

We believe that the slight elevations in the prostatic and bladder cancer odds ratios found in the present study are chance occurrences and are unrelated to the workers' exposure to formaldehyde. The observations leading to this conclusion are as follows:

- The data suggest that by chance the controls chosen for prostate cases were not representative of the underlying control population with respect to formaldehyde exposure. The slight excesses observed for prostate appear to be due to under-representation of exposure in controls, not over-representation of exposure in cases.For example, only 9% of the prostate's controls were exposed whereas 20% of all controls were exposed. The exposure distribution for prostate cases, on the other hand, was similar to that found for all cancer cases and for their controls. Twenty-four percent of prostate cases were exposed vs 21% of all cancer cases.
- Neither the Walrath[3] nor the Marsh study[4] showed an elevated risk for either cancer of the prostate or bladder. Furthermore, the slight excess of prostate cancer observed in the Wong study[5] was limited to workers with the shortest duration exposure, which is contrary to what you would expect to find if formaldehyde were responsible for the excess.
- Our analyses assumed latent periods that were designed to maximize the observed relative risk ratios. This treatment

of latent period may have tended to overestimate the true relative risk.

- The elevations in risk were based on small numbers of cases and were subject to considerable statistical variability. Even so, all of the odds ratios were within the limits of sampling variation and were consistent with a true odds ratio of 1.0.
- The prostate cases came from six plants, the bladder cases from five plants. No one plant showed risks that were particularly higher than the others even though past formaldehyde levels were higher in some plants than in others.
- Prostate and bladder cancers are biologically unlikely sites for the action of formaldehyde, given that the target site in animal studies was restricted to the upper respiratory tract, specifically the nasal cavity.
- The study was subject to the multiple comparison problem and should be interpreted accordingly. That is, any epidemiologic study of this size presents the investigator with a multitude of possible comparisons. A long series of comparisons will, with high probability, result in some comparisons testing significant, even in the absence of any causal relationships.
- In this study the probability that at least one comparison would show a significant difference simply on the basis of chance was better than 40%.

CONCLUSIONS

The data have been analyzed by vital status of the control, source of work history, payclass, tumor site, latent period, duration of exposure, exposure level and frequency, cumulative exposure index, age and year of death, age and year of first exposure, smoking habits, and plant site. In none of these analyses was the relative risk (odds ratio) of cancer for exposed workers significantly greater than 1.0 based on standard statistical analyses.

There was only one cancer death in the category of nose, nasal cavity, or middle ear. This was not a nasal cancer, but rather was a cancer of the middle ear/eustachian tube. Furthermore, the case showed no history of occupational exposure to formaldehyde.

Cancer of the respiratory system showed no signs of being more common in formaldehyde workers than in nonexposed co-workers. The analyses for cancer of the lung were adjusted for cigarette smoking habits, so that the absence of an association between formaldehyde and lung cancer was unlikely to be due to the confounding effects of cigarette smoking.

In short, a thorough analysis of the data suggested that cancer mortality rates in the Company's formaldehyde-exposed workers were no higher than the rates among nonexposed co-workers. These results provide further support for the Company's position that airborne formaldehyde levels of 1 ppm TWA and a 2 ppm ceiling provide adequate worker protection.

ACKNOWLEDGMENTS

T. F. Craggs, Jr., BIOCHEM, Belle; R. D. Richardson, BIOCHEM, Wilmington; R. P. Allen, C&P, Niagara Falls; F. F. Brown, C&P, Grasselli; P. G. Gilby, C&P, Wilmington; K. R. Cortese, ERD, Medical Division; B. W. Karrh, M.D., ERD, Medical Division; M. T. O'Berg, ERD, Medical Division; L. P. Ranken, ERD, Medical Division; W. C. Haaf, F&F, Wilmington; F. K. Hicks, F&F, Parlin; F. B. Peery, F&F, Toledo; E. L. Brennan, PHOTO, Parlin; W. E. Gamble, PHOTO, Rochester; C. C. Griffith, PHOTO, Wilmington; J. F. Doughty, PPD, Washington Works; R. D. Ingalls, PPD, Wilmington.

REFERENCES

1. J. A. Swenberg, W. D. Kerns, et al., Induction of squamous cell carcinomas of the rat nasal cavity by inhalation exposure to formaldehyde vapour. Cancer Res. 40, 3398, (1980).

2. C. F. Reinhardt and N. D. Krivanek, A review of the CIIT final report on the chronic formaldehyde inhalation study and rationale for safe levels of occupational exposure to formaldehyde. Internal Haskell Laboratory publication, 1982.

3. J. Walrath and J. F. Fraumeni, Jr., Proportionate mortality among New York embalmers, in. "Proceedings of the Third CIIT Conference on Toxicology: Formaldehyde Toxicity," J. E. Gibson, ed., Hemisphere, Washington, D.C. (in Press).

4. G. Marsh, Proportionate mortality study of chemical workers exposed to formaldehyde, ibid.

5. O. Wong, An epidemiologic mortality study of a cohort of chemical workers potentially exposed to formaldehyde. ibid.

6. S. Greenland and D. C. Thomas, On the need for the rare disease assumption in case-control studies, Am. J. Epidemiol., 116, 547 (1982).

7. N. E. Breslow and N. E. Day, Statistical Methods in Cancer Research, Volume I. International Agency for Research on Cancer, Scientific Publication No. 32, Lyon, 1980.

8. K. J. Rothman, Induction and latent periods, Am. J. Epidemiol., 114, 253 (1981).

9. K. T. Copeland, H. Checkoway, A. J. McMichael and R. H. Holbrook, Bias due to misclassification in the estimation of relative risk, Am. J. Epidemiol., 105, 488 (1977).

DISCUSSION

DR. BLAIR: (Aaron Blair, National Cancer Institute.) I have a couple of questions for you: 1. What sort of odds ratio could you detect? 2. What was the power for, say, lung cancer, prostate cancer, and colorectal cancer?

MR. FAYERWEATHER: For lung cancer it was about 2. Prostate was 4.4.

DR. BLAIR: The other thing. You made the statement that the only thing that seemed to show a dose effect, although it was not statistically significant, was prostate cancer. Early in your presentation you said that in reviewing other information you concluded it was not related to formaldehyde exposure. Would you elaborate on that and tell us why you drew that conclusion?

MR. FAYERWEATHER: Yes. First of all, we have what we call the multiple comparison problem. We made many, many comparisons, and even if there were no association whatsoever between formaldehyde and cancer, given the number of cancer sites that we looked at and the number of comparisons we made and the different ways that we looked at exposure, the chances that something would show up simply by chance were very high. In fact, they were better than 40 percent in this type of study. So that was one point against causality. There were no statistically significant elevations or trends. Although we found slight excesses for prostate cancer, these were not statistically significant. Neither the Marsh nor the Walrath studies suggested problems with the prostate, although the Wong study did find an excess when he looked at latent periods of over 20 years. When he did subsequent analyses of duration of exposure, his excess incidence would be confined to workers who had shorter duration rather than longer duration of exposure.

When we looked at the exposure rate in the controls for the prostate cancer deaths they did not appear to be representative of the controls for all cancer sites or for controls for other cancer sites. It appears what we did by chance is pick non-representative controls with respect to exposure to formaldehyde.

When we did our analyses, we picked a latent period that would maximize the odds ratio so that we were really loading the dice to begin with.

The excesses of prostate cancer did not correlate with plants having the highest formaldehyde exposures, and finally prostate is a biologically unlikely target of formaldehyde.

DR. BLAIR: A thought occurred to me, and obviously you have analyzed this in many different ways and as thoroughly as you possibly can. When you do the matched pairs analysis you have severe numbers problems; not only with the cases but, also with the controls. Did you lump the controls and do a stratified analysis, which would help you out of the non-representativeness problem of the prostate controls?

MR. FAYERWEATHER: No, I did not.

MR. SILVERSTEIN: (Michael Silverstein, United Auto Workers.) I have two questions, one of which has to do with the power and I think you may have answered this in response to Dr. Blair's question. I am just not sure I heard it correctly. Did you indicate that your results would not be able to preclude a doubling of the risk for respiratory cancer?

MR. FAYERWEATHER: No. We had adequate power to detect a doubling for lung cancer, that is, lung, bronchus and trachea. We could not preclude a doubling, say of prostate.

MR. SILVERSTEIN: Second question, I understand that in your group a number of the cases and perhaps the controls may have been exposed to other known carcinogens, like asbestos and chromates, and I would like you to comment on how this potential confounding may have affected the results, particularly whether the odds ratios could have been substantially diluted by this?

MR. FAYERWEATHER: Most of these other known or suspected carcinogens at the plants, such as asbestos, carbon tetrachloride, chloroform, and benzene were present only in trace amount or in laboratory quantities. However, for the purposes of completeness, you have to say that yes, there were other potential carcinogens at the plants. Now, if the non-formaldehyde workers were exposed to carcinogens which the formaldehyde workers were not exposed to, it is possible their cancer risks would have been elevated which, would have tended to reduce the odds ratios that we observed in this study.

MR. EDWARDS: (Gordon Edwards, Toxicon Associates.) I also have two questions. The first has to do with the fact that about half your cases came from a single plant. I assume that is Belle, West Virginia?

MR. FAYERWEATHER: Yes, it is.

MR. EDWARDS: Which is a plant in an area well known for rather high amounts of pollution, many other chemical plants, and so forth. I wonder how you think that might have affected your results? Clearly in that area there is an elevated incidence of background cancer.

MR. FAYERWEATHER: Yes, we have done a company study of our own of the cancer morality at the Belle plant by itself. This was a cohort study and it was done by Dr. Maureen O'berg. When we compared cancer deaths at Belle Plant versus the rest of the company, certain sites did stand out, such as lung and kidney. When we compared the cancer mortality at Belle relative to the rates in the Kanawha Valley itself, there was no excess of lung cancer. There still was an excess of kidney cancer. Subsequent to that study, NIOSH did their own study. They were not able to identify excesses of kidney or eye cancer that were related to any particular work area at the Belle Plant. What we are doing here is comparing cancer rates in workers at, say, Belle, exposed to formaldehyde versus cancer rates in workers at Belle who are not exposed to formaldehyde. So, it should be a very valid, meaningful comparison.

MR. EDWARDS: My point is that perhaps your should try recasting your data, taking out the Belle workers and seeing what happens.

MR. FAYERWEATHER: I see. We analyzed the data on a company wide basis to see if there was any one particular plant that stood out. In that situation you are looking at all cancer deaths. There was no one particular plant that had a significant excess. If you look at Belle, for all cancer deaths, Belle was actually less than one. The risk in formaldehyde workers at Belle was actually less than the risk in non-formaldehyde workers. This was for all cancer deaths. When you look at lung cancer which was high at Belle compared to the rest of the company but low at Belle compared to the Kanawha Valley, there is nothing that jumps out and looks surprising. So, it does not appear that Belle unduly weighed our results.

MR. EDWARDS: Now, these are all exposed workers, not your C3 group or anything?

DR. FAYERWEATHER: That is right.

MR. EDWARDS: How about smokers in the C3 group?

MR. FAYERWEATHER: I have not looked at that yet.

DR. ACHESON: (Donald Acheson, Medical Research Council in England.) Mr. Fayerweather, you properly pointed out that your material is limited to pensioners. What are the criteria for inclusion on the pension roll, and what proportion of the total exposed population do pensioners represent?

MR. FAYERWEATHER: I don't remember that, Dr. Acheson. I have it in one of my tables, and it will appear in the abstract. It will appear in the paper, but I would have to look it up. I think it is about 50/50, 50 percent pension, 50 percent active.

DR. ACHESON: How long do you have to work in the plant before you are a pensioner?

MR. FAYERWEATHER: You obtain a pension after 10 years of service with the company. People who leave the company prior to 10 years of service would have no pension, and if they subsequently died we would not know of their death.

DR. ACHESON: I think that is an important element of selection. Ten years is a long time. That means you could be exposed five or nine years and not be in the study at all.

MR. FAYERWEATHER: That is right. We have done a quantitative analysis of that, and we found the mortality rates in the people who left without a pension would have had to have been five to ten times higher than what we observed in this study for it to have made any difference in our study results.

DR. KANG: (Dr. Han Kang, OSHA.) In your study, cases came from eight plants and deaths ocurred between 1957 and 1979. Yet two of your plants stopped producing formaldehyde after 1960. How many deaths were there between 1957 and 1960 in these two plants, the Parlin and Washington plants?

MR. FAYERWEATHER: I recently checked that. They represent about 2 percent. It is only about 2 percent because the plants that started producing formaldehyde in 1960 and after only contributed a limited number of deaths to the study. When one does the analyses, since you are actually looking at discordant pairs and since the control was picked from the roster during the same year that the case was at the plant, you would have concordant pairs for those individuals who started, or were first exposed, prior to 1960. Therefore, those individuals would have dropped out of analysis. So, there is no bias there.

DR. KANG: But would that not affect your analysis by 2 percent?

MR. FAYERWEATHER: No. Those individuals will actually drop out of the analysis. When you look at discordant pairs there is no way that those individuals could be discordant with their controls. So they would drop out of the analysis.

Now, the sample size calculations should actually have been based on 98 percent of the 481, rather than 100 percent of 481. That might make a decimal point or two difference in the detectable odds ratio, but it would have no substantial effect.

DR. HECK: (Henry Heck, CIIT) I just want to support what you say about the prostatic cancer data. We have analyzed formaldehyde in the blood of rats exposed to 15 parts per million of formaldehyde for 2 hours, and the blood formaldehyde concentration in the exposed rats was 0.0749 micromoles per gram of blood. In control rats it is 0.0746 micromoles per gram of blood, and the standard deviation is 0.0067 micromoles per gram in both cases.

MR. FAYERWEATHER: I would like to present something additional. A number of people have commented in the past and said, "Gee, didn't you overmatch? You matched on age, sex, pay class, years of service, etc." Taking, for example, they feel we should not have matched on pay class, since if a case was wage and the control was also wage, they felt that would guarantee that both the case and the control would have had formaldehyde exposure.

The answer is no. After those questions were posed, I went back and analyzed the data according to pay class. I would like to present one slide here (Ed. Note: Table 8 above). In the first third of the Table we are looking just at wage roll employees. All the cases are wage roll and all the controls are wage roll. You see that about 21 percent of this group had potential exposure to formaldehyde. The odds ratio is about 0.88.

If you go to the middle third of the Table, the salary roll, all the cases and controls were salaried. The exposure rate was 19 percent, very close to what it was with wage roll cases. The odds ratio again is very close to what it was for the wage roll case - 0.73. As I have already shown you, we combined wage and salary rolls, which gave a total percent exposed of 21 percent and an odds ratio of 0.85. So, on the basis of pay class, there is no reason to say that we had overmatching. If there were overmatching, you would not find as many discordant pairs. If you look at wage roll exposed controls, you have 51 cases that were not exposed and 51 controls that were exposed. Now, if we were overmatching, you would not tend to find this many discordant pairs.

If you look at "exposed wage roll cases" vs. "not-exposed wage roll control," 45 cases exposed and 45 controls not exposed. Again, you would not tend to find that many discordant pairs, if you are having problems with overmatching.

APPENDIX A

FORMALDEHYDE STUDY CODE SHEET FOR WORK HISTORIES

S.S. No: ___ ___ ___-___ ___-___ ___ ___ ___ Case No.: ___ ___ ___-___

Birthdate (mo/yr): ___ ___ / ___ ___

Date Hired (mo/yr): ___ ___ / ___ ___

Department Code

Plant Site Code

Initial Payclass (1=wage, 2=salary): ___

Final Payclass (1=wage, 2=salary): ___

Number of Jobs Listed Below: ___ ___

List chronologically all jobs:

C = Continuous-direct (C1, C2, or C3)
I = Intermittent (I1 or I2)
B = Background

Job	Area and Job Name	Area Code	Job Code	Date in (mo/yr)	Date out (mo/yr)	Exposure Category
1.						
2.						
3.						
4.						
5.						
6.						
7.						
8.						
9.						
10.						
11.						
12.						
13.						
14.						
15.						
16.						
17.						
18.						

Last day at plant (mo/yr) ___ ___/ ___ ___

Termination Reason: 1 Still Active; 2 Pensioned; 3 Resigned; 4 Deceased; 5 T & P Disability; 6 Transferred; 7 Other

Worker's name
(last, first, middle initial): ___ ___________ ______________________

APPENDIX B

SMOKING HISTORY QUESTIONNAIRE

Subject's name: ____
Last First Middle initial

What is your relationship to the subject? ______________________________

____

The smoking history given below is based primarily on:

a. Your memory b. Medical records

If history is based primarily on your memory, how many years had you known the subject?

a. < 1 year
b. 1-5 years
c. 6-10 years
d. 11-15 years
e. > 15 years

When was the last time (e.g., 1968) that you saw the subject? 19___

With respect to <u>cigarettes</u> the subject was a:

a. Current smoker
b. Ex-smoker
c. Never smoked (never smoked as much as one cigarette per day for as long as one year)

If the subject was a current or ex-smoker:

When the subject smoked, he was a:

a. Light smoker
b. Moderate smoker
c. Heavy smoker
d. Don't know

How many packs per day did he smoke when he was a smoker?

a. Less than 1/2
b. 1/2 - 1
c. 1 - 2
d. 2 or more
e. Don't know

If he was an ex-smoker, in what year did he quit smoking? 19___

With respect to <u>pipes</u>, the subject was a:

a. Current smoker
b. Ex-smoker
c. Never smoker
d. Don't know

With respect to <u>cigars</u>, the subject was a:

a. Current smoker
b. Ex-smoker
c. Never smoked
d. Don't know

Name of person who filled out the above questionnaire.

(Please print): ____

APPENDIX C

MATERIALS THAT HAVE UNDERGONE RISK ASSESSMENT BY HASKELL LABORATORY AND WHICH ARE FOUND AT THE FORMALDEHYDE STUDY PLANTS

<u>Washington Works (PPD)</u>
Asbestos
Benzene*
Carbon tetrachloride*
Chloroform*
4-Dimethylaminobenzene
4-Aminodiphenyl
Dimethyl sulfate
p-Dioxane
Lead chromate
Tetrachloroethylene
B-propiolactone
Trichlorethylene
Zinc chromate
Freon-22

<u>Parlin (F&F)</u>
Acrylonitrile
Asbestos
Benzene*
Propyleneimine
Carbon tetrachloride*
Chloroform*
Dioxane
Lead chromate
Lead molybdate
Strontium chromate
Trichloroethylene
Zinc chromate
Cadmium
Fuel oils

<u>Belle (BIOCHEM)</u>
Asbestos
Benzene
Dimethylnitrosamine*
Carbon tetrachloride*
Benomyl
MBC
Chloroform*
Dimethylacetamide
Dimethyl formamide
Dimethyl sulfate
1,4-Dioxane

<u>Toledo (F&F)</u>

Acrylonitrile
Asbestos
Benzene*
Lead chromate
Lead molybdate
Strontium chromate
Zinc chromate
Cadmium
Fuel oils

<u>Toledo (C&P)</u>
None

<u>Parlin (PHOTO)</u>
Asbestos
Chloroform
Trichloroethylene
Tetrachloroethylene
ETU
Thiourea*
Benzene*

<u>Rochester (PHOTO)</u>
Asbestos
Acrylonitrile*
3-amino-1,2,4-triazole
Ammonium perfluorooctanoate (C-8)
Carbon tetrachloride*
Freon-22
Chloroform*
Dimethyl formamide*
o-Dianisidine*
1,1-dimethyl hydrazine*
Dimethyl sulfate*
Dioxane*
Epichlorhydrin*
2-ethoxy ethanol
ETU*
Formamide*
Hydrazine*
Lead nitrate
Lead
Light green S-F dye
Tetramethyl thiourea*
Thioacetamide
Thiourea*
Toluene-2,4-diamine
o-Toluidine*

*Present as impurity in trace amounts, used in laboratory, or used in R&D.

CHAPTER 4

MORTALITY OF ONTARIO UNDERTAKERS: A FIRST REPORT

R. J. Levine, D. A. Andjelkovich, L. K. Shaw, and R. D. DalCorso

Chemical Industry Institute of Toxicology
Research Triangle Park, North Carolina

In order to examine the mortality of a cohort of workers exposed to formaldehyde, the experience of male undertakers first licensed in Ontario during 1928-1957 was determined as of January 1, 1978. Of 1477 persons enrolled in the cohort, death certificates were obtained for 337 who died during the period of observation; 923 are known to have survived. Efforts to gather information about 217 persons (15%) currently lost to follow-up are under way.

Because death rates for Canadian men were not yet available in suitable computer format, the U.S. white male population was used. It has a similar mortality experience as the standard for this preliminary report. Sites thought to be at special risk of cancer from exposure to formaldehyde -- skin, nasal passages, buccal cavity, pharynx, and larynx -- were of particular interest. No deaths due to nasal cancer were observed. Mortality attributed to cancers of the buccal cavity, pharynx, and respiratory system exclusive of trachea, bronchus, and lung was less than expected (2 deaths observed, 3.7 expected, SMR = 54). Cirrhosis of the liver was the only cause of death found to be significantly in excess (SMR = 172).

INTRODUCTION

Because of the nasal cancers observed in rodents exposed to formaldehyde gas, attention has focused on the possibility that exposure to formaldehyde might cause cancer in humans. A number of agents found in the workplace -- nickel, chromium, mustard gas, isopropyl oil, radium, polycyclic aromatic hydrocarbons, and wood dust -- have been linked to cancer of the nose and paranasal sinuses.[1] To date, however, formaldehyde has not been implicated specifically. While rodents are obligatory nose-breathers, humans may breathe through the nose or mouth and human skin is unprotected by fur. It would seem, therefore, that in people, sites at special risk due to contact with formaldehyde would be skin, nasal passages, buccal cavity and pharynx, larynx, and possibly esophagus and lung.

The mortality experience of several occupational groups with known exposure to formaldehyde has been evaluated -- that of pathologists, morticians, and chemical workers.[2-5] With the exception of an elevated proportion of deaths due to skin cancer among New York undertakers, increased cancer mortality has not been observed at the sites of special risk from formaldehyde gas. Excesses of cancer of the brain, kidney, prostate, and lymphatic and hematopoietic tissues have been noted. This report will present preliminary findings of a retrospective cohort mortality study of undertakers in Ontario, Canada.

Embalmers disinfect, preserve, and restore dead bodies in preparation for funeral. During the course of work they are exposed principally to the toxic effects of formaldehyde and its polymers, which constitute the primary active ingredients of embalming fluids, jellies, and powders.

Mean time-weighted-average concentrations of formaldehyde gas were recently assayed from the breathing zone of embalmers at six West Virginia (U.S.) funeral homes. Levels of 0.3 and 0.9 ppm, respectively, were measured during the approximately one hour required to embalm an intact body and three hours required for an autopsy case. Peak half-hour concentrations were 0.4 and 2.1 ppm.[6] A survey conducted in West Virginia during 1979 revealed that on the average, during years in which they embalmed, undertakers prepared approximately 75 bodies per year, including 15 that had been autopsied.[7]

Not all undertakers embalm, but almost all have done so at some time during their careers and will have been exposed to formaldehyde. Examination of the experience of undertakers, therefore, is valuable in assessing the possible effects of formaldehyde exposure on human health.

Since 1914 the names of undertakers who qualified for licensure in Ontario, Canada are listed in annual directories published by the Ontario Board of Funeral Services. From these directories a cohort of 1477 men first licensed by examination of the board during 1928-1957 inclusively, was identified for a retrospective cohort mortality study.

METHODS

The following information was extracted from board records for each cohort member: name in full; date of birth; date when last license fee or payment for continuing education was received; town and province of business; town and province of residence; and, when known, year or date of death. In addition, for those first licensed in 1938 or subsequent years, town and province of birth were obtained; and for licensees after 1948, father's name.

The mortality of the cohort was followed from first licensure through December 31, 1977. Persons were considered alive at the closing date of the study if records of the Ontario Board of Funeral Services or the Registrar of Motor Vehicles indicated that a fee had been paid or a communication received on or after January 1, 1978. A list of persons known to be dead or whose vital status was unknown was submitted together with identifying information to the Vital Statistics and Disease Registries Section of Statistics Canada. Deaths of Canadian residents during 1928-1949 were ascertained manually from a search of fiche files and for the period 1950-1977 through a computerized system of record linkage.[8] Persons known to have died during World War II were also sought through the Canadian War Graves Commission.

Of the 1477 cohort members, thus far 931 (63%) are known to have been alive and 337 (23%) dead on January 1, 1978. The vital status of 209 persons (14%) is unknown. Men of unknown vital status contributed person-years of observation until the date they were last known to be alive, usually the date of their last payment to the Ontario Board of Funeral Services.

Death certificates were obtained for all but six of the deceased. Underlying and contributing causes of death were coded by trained nosologists of Statistics Canada, according to the eighth revision of the International Classification of Diseases adapted for use in the United States. The six individuals from whom death certificates were not found had been killed in the war and were assigned E995 (injury due to war operations by other and unspecified forms of conventional warfare) as the underlying cause of death.

Numbers of observed and expected deaths and the ratio of observed to expected expressed as a percentage of "standardized mortality ratio" (SMR) were determined for each cause. Expected deaths were computed by applying age and calendar year-specific mortality rates of Ontario men to the experience of the cohort, using a computer program developed by Dr. Richard Monson.[9] Statistical significance was evaluated through the same program, which determined chi square with one degree of freedom. The cumulative Poisson probability was calculated separately whenever expected deaths were less than five. Because of the lack of computer tape incorporating Ontario rates for all causes of interest and because rates prior to 1950 were not available to us, the mortality schedule of U.S. white males was used as a supplement. While overall the experience of Ontario men is not dissimilar to that of U.S. white males, mortality rates among Canadians tend to be lower.

Person-years of observation in this report total 44,306, of which 16,422 occurred twenty or more years since first licensure. On average, each member of the cohort contributed 30 years of observation, was 26 years old at admission to the study, and if deceased during the study period, was 57 years old at death.

RESULTS

Results for selected malignant and non-malignant diseases are given in Tables 1-3. Cancers of special interest were those at sites of possible contact with formaldehyde gas and those of the brain, kidney, prostate, and lymphatic and hematopoietic tissues, which had been found in excess in studies of other occupational groups exposed to formaldehyde.

SMRs (U.S.) in Tables 1 and 2, where mortality rates of U.S. white males were used to compute the number of expected deaths, span the entire period of observation from 1928 through 1977. Whenever Ontario rates were used (Tables 1 and 3), on the other hand, observation begins with 1950 since rates for preceding years were not available.

Mortality due to all cancers combined was not elevated, regardless of whether U.S. or Ontario rates were used; and a significant increase in mortality was not detected for any site-specific cancer. Cancer of the nose was neither an underlying nor a contributing cause of any death. Deaths from cancers at other sites of potential contact with formaldehyde gas - skin, buccal cavity and pharynx, larynx, lung, and esophagus - were less than expected, as were deaths from cancers of the kidney and prostate. Small excesses of cancer of the brain and

TABLE 1

Cancers of Special Interest with Respect to Formaldehyde Exposure

CANCER	O/E (U.S.)	SMR (U.S.)	O/E (Ontario)	SMR (Ontario)
Skin	0/1.3	0	---*	---*
Nose**	0/0.3	0	0/0.2	0
Buccal cavity & pharynx	1/2.4	42	---*	---*
Larynx	1/1.1	93	1/0.9	112
Lung	19/21.3	89	18/18.3	98
Esophagus	0/1.6	0	0/1.6	0
Brain	3/2.5	118	3/2.5	122
Kidney	1/1.8	56	1/1.5	66
Prostate	2/3.1	65	---*	---*
All lymphatic and hematopoietic	9/7.2	125	8/5.6	134
All cancers	60/67.7	89	56/60.7	92
All causes	337/ 370.0	91	309/ 294.5	105

* Not available.
**ICD 160 & 163.

O/E (U.S.) - Observed/expected deaths during 1928-1977; mortality rates of U.S. white males were used for computing expected deaths.
O/E (Ontario) - Observed/expected deaths during 1950-1977: mortality rates of Ontario males were used for computing expected deaths.
SMR - Standardized mortality ratio.

TABLE 2

Cancers of Special Interest with Respect to Formaldehyde Exposure
Latency Analysis

	10 Year Latency		20 Year Latency	
CANCER	O	SMR (U.S.)	O	SMR (U.S.)
Skin	0	0	0	0
Nose*	0	0	0	0
Buccal cavity & pharynx	1	44	1	51
Larynx	1	96	1	110
Lung	19	91	18	97
Esophagus	0	0	0	0
Brain	3	133	3	179
Kidney	1	59	1	70
Prostate	2	66	2	71
All lymphatic and hematopoietic	7	109	6	119
All cancers	58	91	52	96

*ICD 160 & 163.
O - Observed deaths (1928-1977).
SMR - Standardized mortality ratio.

of the lymphatic and hematopoietic tissues were noted. In general, these observations remained unchanged if observed and expected deaths during the first 10 or 20 years after licensure were excluded to allow for a period of latency (Table 2). Non-significant increases in SMRs were noted for brain cancer, but these were based on fewer than 2.5 expected deaths.

As can be seen in Table 3, mortality from all causes combined was somewhat greater than expected as the result of a significant

TABLE 3

Non-malignant Diseases

CAUSE	O/E (Ontario)	SMR (Ontario)
Infectious & Parasitic	2/ 2.9	70
Tuberculosis	1/ 1.6	64
Circulatory System	173/ 151.4	114
Ischemic Heart Disease	129/ 113.5	114
Respiratory System	13/ 15.8	82
Influenza & Pneumonia	9/ 6.6	136
Bronchitis, Emphysema, Asthma	3/ 6.2	48
Digestive System	27/ 13.7	197**
All Non-malignant Diseases	228/ 220.2	114*
All Causes	309/ 294.5	105

*$p < 0.05$
**$p < 0.001$
O/E (Ontario) - Observed/expected deaths during 1950-1977; mortality rates of Ontario males were used for computing expected deaths.
SMR - Standardized mortality ratio.

increase in mortality due to non-malignant diseases. This may be attributed to a small excess of circulatory system deaths (SMR 114), which comprise the majority of deaths due to non-malignant diseases, and a twofold excess of deaths related to the digestive system. The preponderance of digestive system deaths was from cirrhosis of the liver (18), other or unspecified liver disease (1), and acute pancreatitis (2), diseases frequently associated with alcoholism. The increase in these deaths is highly significant ($p < 0.001$). It should be noted that 8 of the 18 deaths from cirrhosis of the liver were specified to be of alcoholic origin (ICD 571.0). Mortality due to infectious and parasitic diseases or to chronic diseases of the respiratory

TABLE 4

Moderate or Substantial Excesses of Mortality from Specific Diseases*

DISEASE	O/E (U.S.)	SMR (U.S.)	O/E (Ontario)	SMR (Ontario)
Liver Cancer	2/1.4	145	2/0.5	389
Cirrhosis of the Liver	18/10.5	171**	---[+]	---[+]
Leukemia & Aleukemia	5/2.9	175	4/2.3	175
Chronic Rheumatic Heart	8/5.4	150	---[+]	---[+]

* Diseases with 2 or more observed deaths and SMRs $\geq$ 150.
**$p < 0.05$.
[+] Not available.
O/E (U.S.) - Observed/expected deaths during 1928-1977: mortality rates of U.S. white males were used for computing expected deaths.
O/E (Ontario) - Observed/expected deaths during 1950-1977: mortality rates of Ontario males were used for computing expected deaths.
SMR - Standardized mortality ratio.

system (bronchitis, emphysema, asthma) was less than expected, while deaths ascribed to acute respiratory diseases (pneumonia, influenza) were increased.

Moderate or substantial excesses of mortality from specific diseases are recorded in Table 4. This table includes all diseases for which there were SMRs (U.S. or Ontario) of 150 or greater and 2 or more observed deaths, based on the entire experience of the cohort. The greatest SMR (389) was for liver cancer when Ontario rates were used. Deaths from cirrhosis of the liver were significantly increased. A significant excess of deaths attributed to chronic rheumatic heart disease was observed after excluding the first 20 years of licensure: SMR (U.S.) 254, $p < 0.05$.

DISCUSSION

The preeminent feature of the mortality experience of this cohort is the highly significant excess of deaths due to non-malignant diseases of the digestive system. This indicates that the study group may have consumed more alcoholic beverages than the general population; and it might account for the increases in mortality from liver cancer, diseases of the circulatory system and acute respiratory diseases, which are characteristically elevated among alcoholics.[10]

Undertakers must frequently handle bodies of persons who have died with infectious diseases. Mortality due to these causes, nevertheless, was not increased. Deaths attributed to chronic rheumatic heart disease were in excess, especially after the first 20 years of licensure had been excluded. While this could indicate a greater risk of streptococcal infections and rheumatic fever, 7 of the 8 deaths were from valvular heart diseases not specified as rheumatic and might possibly have originated from other causes.

It has been suggested that exposure to formaldehyde may augment the frequency of asthma and acute and chronic bronchitis[11,12] and, over the long-term, decrease expiratory flow rates.[13] Ontario undertakers, however, did not have an excess of deaths from chronic respiratory diseases.

Studies have been conducted of other occupational groups exposed to formaldehyde. The mortality experience of cohorts of British pathologists and formaldehyde-exposed workers at a U.S. chemical plant revealed excesses of lymphatic and hematopoietic neoplasms and prostatic cancer, respectively.[2,5] Proportional mortality from cancers of the skin, kidney, and brain was elevated among New York undertakers.[3] Mortality due to these causes was not increased in this study. Except for skin, none of the sites mentioned above was suspected _a priori_ to be at risk from formaldehyde exposure. Thus, while these data may be employed to generate hypotheses for testing, they cannot be used to substantiate a relationship between formaldehyde and cancer. The biological plausibility of such a relationship at these locations, furthermore, is in doubt. The skin and upper respiratory tract receive by far the greatest exposure to formaldehyde gas; and it is dubious whether an increase in concentration of formaldehyde or its metabolites could be detected in the internal organs, even at levels far in excess of usual human exposure. (In rats exposed for two hours to 15 ppm formaldehyde and in unexposed controls the concentrations of formaldehyde in the blood appeared to be identical).[14] If formaldehyde were to affect human health, therefore, it should exert its influence at the sites of contact. Ingestion of alcohol has been shown to increase the risk of several cancers, including those of the buccal cavity, pharynx, larynx, and esophagus.[15] Despite apparent heavy use of alcoholic beverages by the study cohort, excess deaths were not observed from cancers at sites of potential contact with formaldehyde gas including those at increased risk from the consumption of alcohol. Thus the likelihood is enhanced that formaldehyde exposure among Ontario undertakers has had no effect on cancer mortality.

It should be noted that these results are to be viewed as preliminary, since follow-up of the cohort is not yet complete. Continuing efforts are being made to ascertain at the closing date of the study the vital status of as many members of the cohort as possible.

ACKNOWLEDGMENT

We gratefully acknowledge the considerable assistance of Donald B. Steenson, Registrar of the Ontario Board of Funeral Services, and Martha E. Smith, Chief of the Occupational and Evnironmental Health Research Unit, Vital Statistics and Disease Registries Section, Health Division, Statistics Canada.

REFERENCES

1. G. C. Rousch, J. W. Meigs, J. Kelly, J. T. Flannery, H. Burdo, Sinonasal cancer and occupation: A case-control study, Am. J. Epidemiol., 111, 183, (1980).

2. J. M. Harrington, H. S. Shannon, Mortality study of pathologists and medical laboratory technicians, Br. Med. J., 4, 329 (1975).

3. J. Walrath, J. F. Fraumeni Jr., Proportionate Mortality among New York Embalmers. In "Proceedings of the Third CIIT Conference in Toxicology: Formaldehyde Toxicity", J. E. Gibson, ed., Hemisphere, Washington, D.C. (In Press).

4. G. M. Marsh, Proportional mortality patterns among chemical plant workers exposed to formaldehye, Br. J. Ind. Med., 39, 313 (1982).

5. O. Wong, An Epidemiologic Mortality Study of a Cohort of Chemical Workers Potentially Exposed to Formaldehyde, with a Discussion on SMR and PMR. In "Proceedings of the Third CIIT Conference in Toxicology: Formaldehyde Toxicity", J. E. Gibson, ed., Hemisphere, Washington D.C. (In Press).

6. T. R. Williams, R. J. Levine, P. B. Blunden, Chemical exposure of embalmers. Occupational Health Studies Group, University of North Carolina, Chapel Hill, NC (1980).

7. R. J. Levine, R. D. DalCorso, P. B. Blunden, M. C. Battigelli, The Effects of Occupational Exposure on the Respiratory Health of West Virginia Morticians. In "Proceedings of the Third CIIT Conference in Toxicology: Formaldehyde Toxicity", J. E. Gibson, ed., Hemisphere, Washington D.C. (In Press).

8. M. E. Smith, H. B. Newcombe, Automated follow-up facilities in Canada for monitoring delayed health effects, Am. J. Public Health, 70, 1261 (1980).

9. R. R. Monson, Analysis of relative survival and proportional mortality, Comput. Biomed. Res., 7, 325 (1974).

10. W. Schmidt, J. De Lint, Causes of death of alcoholics, Q. J. Stud. Alcohol, 33, 171 (1972).

11. D. J. Hendrick, D. J. Lane, Occupational formalin asthma, Br. J. Ind. Med. 34, 11 (1977).

12. E. R. Plunkett, T. Barbela, Are embalmers at high risk?, Am. Ind. Hyg. Assoc. J., 38, 61 (1977).

13. J. B. Schoenberg, C. A. Mitchell, Airway disease caused by phenolic (phenolformaldehyde) resin exposure, Arch. Environ. Health, 30, 574 (1975).

14. H. d'A. Heck, personal communication, Chemical Industry Institute of Toxicology, Research Triangle Park, NC 27709, (1982).

15. A. J. Tuyns, Alcohol. In "Cancer Epidemiology and Prevention", D. Schottenfeld and J. F. Fraumeni, Jr., eds., W. B. Saunders, Philadelphia, (1982).

DISCUSSION

DR. YODAIKEN: (Ralph Yodaiken, NIOSH.) Your paper was excellent, and I congratulate you on it. Your results are extraordinarily interesting. I think it is an assumption, however, to believe that the deaths from cirrhosis are necessarily alcohol-related. You may have picked up some underlying cause which may be very significant. These people, first of all, do absorb large quantities of formaldehyde, and that is metabolized in the liver. So it may be that we have a cause of cirrhosis that we have not thought about. Secondly, at least in some of your figures, you have suggested that there is an increased incidence of acute infection or a proneness to develop influenza, among other diseases. Now, there might ostensibly be a proneness to infectious hepatitis or something of that nature. Thirdly, I wonder if you have any intention of following up on life styles of embalmers?

DR. LEVINE: First of all, when I was looking around for a group of embalmers to study, I called every state in the Union before I wound up with Ontario, Canada. In conversations with the people who were on the Boards of Funeral Services in the various states, it was mentioned to me several times that embalmers are known to drink. This was also mentioned to me by people at the Ontario Board of Funeral Services. One of them told me that his mother was very concerned when he became an embalmer because she said, "They drink." Dr. Budlovsky, who is here, is with the Ontario Ministry of Labor, and he has some interesting results on studies he has carried out on embalmers in Ontario. He tells me that it is his understanding that a cup of tea, with lemon and a shot of rum is a very good antidote for all kinds of infectious diseases. If embalmers were concerned about infectious diseases on the job, perhaps this would explain an increased prevalence of alcoholism.

However, in regard to the cirrhosis of the liver, I would be principally concerned about, as you pointed out, the possibility that infectious hepatitis through a larger exposure to blood and secretions might be a cause of excess risk. I don't think that I am going to pursue this, but this could be something that might be appropriate to do. In addition, of course, it might be worthwhile to pursue the chronic rheumatic heart disease deaths; really, valvular heart diseases, for which the etiology has not been specified. The possible relationship of this to rheumatic fever and group A beta hemolytic streptococcal infections could possibly be pursued in a case controlled study, using hospital records.

I am not at all concerned about the possibility that formaldehyde, or other agents that embalmers are exposed to, could have had some role in the etiology of cirrhosis of the liver. As far as the formaldehyde goes, results will be presented later that will show that it is metabolized very rapidly at the sites of contact. As Dr. Heck pointed out, there is no evidence for any increase in blood levels of formaldehyde. So, how it could be transported far from the site of origin and do damage in an organ away from the site of contact. I don't understand how that could be of any concern.

DR. YODAIKEN: Can I just follow up that question for one second? You have treated it lightly, but I think it is a little more important than that. I am a pathologist of 20 years' experience, and I think, on an anecdotal basis, that all of the embalmers or morticians that I have met appear to be abstemious individuals. They may be closet drinkers, but it is my personal experience, as well as yours and that of other people, that they do not overindulge. But I think it is worthwhile to follow that point up just to make sure that, in fact, you are dealing with a non-abstemious population.

DR. LEVINE: We had about 11 individuals including eight for whom cirrhosis of the liver was mentioned as the underlying cause of death and alcoholism was mentioned as a contributing cause. Now, I don't know how much credence one should give that. Perhaps any doctor that signs out a death certificate due to cirrhosis of the liver is apt to, with little justification, record alcoholism as a contributing cause.

DR. FESTA: (John Festa, American Paper Institute.) Where do you draw the line when the number of observed and expected deaths are so small that it is not meaningful to calculate an SMR? Is there some standard?

DR. LEVINE: If I were an embalmer, I would not draw that line at all. I would be interested in the mortality experience. I think there are two take-home lessons from a study like this. First of all, we can look at what actually has happened up to now in this group. If there has been no excess of certain causes of death that we are interested in, I think we should take some relief in knowing that. The second take-home lesson from a study of this type is what we can extrapolate from these results to other populations and to this population in the future as to the possible effects of formaldehyde exposure. For these extrapolations, power is a necessary item to consider.

We calculated power for cancers at the sites of contact or possible contact with formaldehyde. Those were the four or five that I mentioned. We had an 80 percent power of detecting a threefold increase for the whole cohort. If you only look at after 20 years have elapsed since first licensure, we had a 80 percent power of detecting a fivefold increase. So, we would only have detected a very large increase, insofar as being able to extrapolate to other populations. We could only say that, "Well, formaldehyde doesn't appear to cause a very large increase in this group for these cancers." Insofar as embalmers are concerned, in particular those in Ontario, I think they should take some sustenance from our findings which did not indicate anything of concern with respect to formaldehyde.

MS. STROUP (Nancy Stroup, Yale University.) I know you don't have underlying rates, but I am wondering if you have any information on histologic type for brain cancer and leukemia?

DR. LEVINE: We only looked at death certificates.

MS. STROUP: But sometimes it is listed on the death certificate.

DR. LEVINE: No, I don't have any information.

DR. STERNBERG: (Stephen Sternberg, Memorial Cancer Center, Sloan Kettering.) I just want to make one comment concerning the liver and cirrhosis. I believe that in the absence of any information regarding a hepatotoxic effect of formaldehyde, you cannot postulate any etiology for cirrhosis due to formaldehyde.

There are practically no substances that produce liver cancer that don't produce hepatotoxicity as a precursor. It was true, as I understand it, that there was no acute liver damage in the rat study.

DR. LEVINE: Thank you very much. Another point: About half of these cases of cirrhosis were signed out as Laennec's cirrhosis, which is singular to alcoholism.

DR. BLAIR: (Aaron Blair, National Cancer Institute.) In regard to overall SMR, the "all-causes" SMR for the group, which was roughly around one; it varied as to which comparison population was used. What sort of interpretation do you put on that, in light of the fact that it usually is considerably lower in a working population and it drops even further as you go up the socioeconomic scale? I would not have been surprised to see this study have an overall SMR of 60, not 100.

DR. LEVINE: Yes, I think these people do have an increased risk of death overall. That is attributed to the increase in non-malignant diseases which we observed. The two sources or principal contributors to this increase were non-malignant diseases of the digestive system and ischemic heart disease, also found to be elevated in populations of alcoholics.

MR. JURINSKI: (Neal Jurinski, Nuchemco.) Do you have any information on your population regarding the usual use of technical assistants, who might receive the major part of the exposure? Do you have any data in this regard?

DR. LEVINE: In a study such as this you can never separate out those individuals who send for the flowers from the ones who do the embalming. So all we know is that, among this group of people, there are individuals who have received considerably more formaldehyde exposure than the general population.

MR. EDWARDS: (Gordon Edwards, Toxicon Associates.) It may be a bit off the subject, but one delivery system I can think of for delivering formaldehyde to internal organs might be the worker exposed to methanol who then metabolizes it to formaldehyde. That is not the subject of this conference, I understand.

CHAPTER 5

SKIN INITIATION/PROMOTION STUDY WITH FORMALDEHYDE IN SENCAR MICE

F. Spangler and J. M. Ward

Microbiologial Associates
Bethesda, Maryland
and
National Cancer Institute
Bethesda, Maryland

Certain groups of people are potentially exposed topically to formaldehyde (HCHO). We became concerned with the possibility that this chemical might play a role in the development of skin cancer in these people. It was decided to test HCHO for its potential to induce papillomas on the mouse backskin under conditions that would elucidate whether HCHO was acting as an initiator (I), a promoter of cancer (P), as a complete carcinogen (CC), or as a non-carcinogen. We chose as our test model the SENCAR mouse (SENsitive to two-stage skin CARcinogenesis). The preliminary data from the first 48 wks. of an ongoing 78 wk. study are reported.

The vehicle for all test compounds was acetone (A), HCHO was used at 3.7%-4.0% in A. DMBA at 20μg/dose, and TPA at 1.25 μg/dose. I's were applied only once; P's were applied either once or twice/wk. Each test group contained 30 female SENCARS.

To date, the data indicate that HCHO probably is not a CC. The data also indicate that HCHO probably is not an I. The data on HCHO as a promoter are inconclusive. HCHO may or may not be a weak promoter of skin carcinogenesis in the SENCAR. When terminated, the study will be fully evaluated histopathologically and statistically.

TABLE 1

Experimental Design for the Study of Skin Tumor Initiation and/or Promotion with Formaldehyde

Group	Treatment*	Purpose of Group
1	F,F	F as a potential CC
2	F,T	F as a potential I
3	D,F	F as a potential P
4A1	D,A,T	Positive control
4A2	D,A,A	Negative control
4B	F,A	Negative control
4C	A,T	"Negative" control
4D	A,A	Negative control
4E	A,F	Negative control

*Treatment Legend: F = Formaldehyde
T = 12-O-Tetradeconoyl-phorbol-13-acetate
D = Dimethylbenzanthracene
A = Acetone

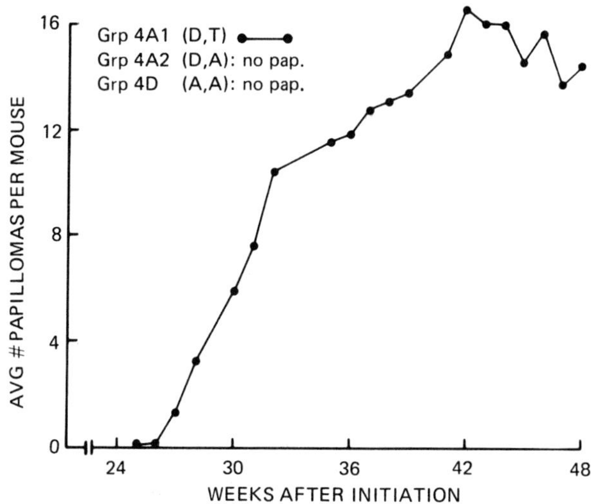

Fig. 1. Incidence of skin papillomas in SENCAR mice following application to the skin of DMBA (D), TPA (T), or ACETONE (A) in the indicated manner. DMBA or acetone alone, or in combination, did not induce skin papillomas.

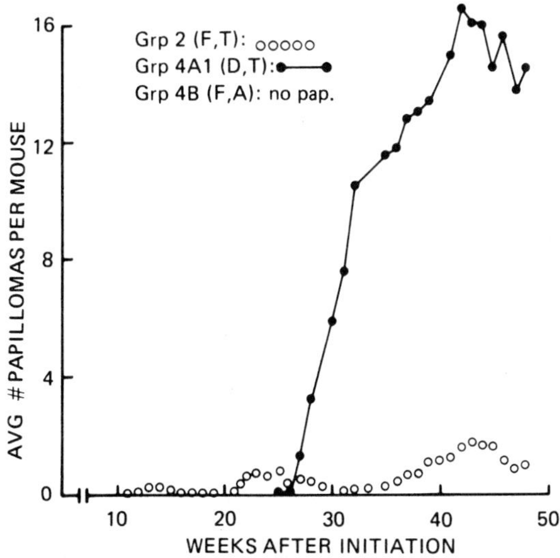

Fig. 2. Incidence of skin papillomas in SENCAR mice following skin application of formaldehyde and TPA, DMBA and TPA, or formaldehyde and acetone. DMBA in combination with TPA was much more effective than other treatments in inducing papillomas.

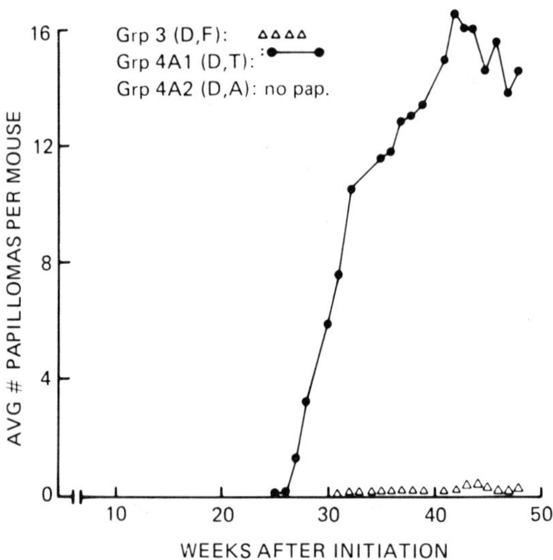

Fig. 3. Incidence of skin papillomas in SENCAR mice following application to the skin of DMBA and formaldehyde. These applications were compared to groups treated with DMBA and TPA or DMBA and acetone. Only the DMBA and TPA group gave a notable response.

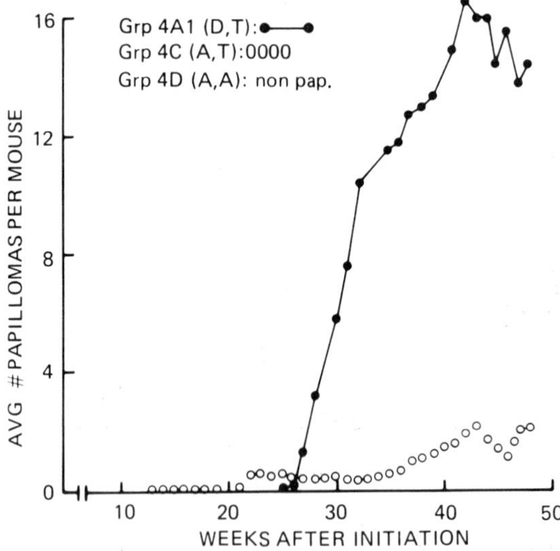

Fig. 4. Incidence of skin papillomas in SENCAR mice following application to the skin of DMBA and TPA, acetone and TPA or acetone and acetone. The incidence of papillomas following acetone and TPA demonstrates a background of initiation in these mice.

In his introductory remarks Dr. Spangler* indicated that the purpose of their study was to investigate the carcinogenic potential of formaldehyde when applied to the skin of mice. He then explained that their study was not a classical chronic skin application study. He also explained that their study was not a study to determine mechanisms. Instead, their design was intended to evaluate the potential of formaldehyde to initiate cancer, to promote cancer or to act as a complete carcinogen.

Dr. Spangler also discussed the reasons that led to the selection of the female SENCAR (SENsitive to skin CARcinogens) mouse for their study. He pointed out that this strain has been shown to be a good model for two-stage (initiation-promotion) carcinogenesis.

Dr. Spangler also explained why they had selected acetone as a solvent. He mentioned its wetting and spreading properties, which help assure distribution of the solute over the skin of the mice; its volatility, which leads to good contact of the residual solute with the skin of the test animal; and the uneventful

---

*Editor's Note: Drs. Spangler and Ward did not submit a manuscript for the Proceedings. They did agree to publication of the Abstract of their paper, a table showing the study design and the figures they presented. The latter show most of their results at the time of the Conference.

To supplement this information, the Editors prepared this report of their understanding of Dr. Spangler's presentation at the Conference. It was prepared by the Editors in order to provide the reader with a more complete account of the paper than is given by the Abstract, the table and the figures.

history of the use of acetone as a solvent for skin application carcinogenicity studies.

As further background information, Dr. Spangler also reviewed the initiation-promotion theory of carcinogenesis. He pointed out that according to this theory, the first step in the development of a cancerous cell is the reaction of the genetic material of a normal cell with an "initiating" agent. This step is not believed to be reversible. If it is not counteracted and is not lethal to the cell, the cell will remain initiated and viable for its normal lifespan. This theory also postulates that the cell will not develop into a true tumor cell until after a subsequent intervention of some sort (promotion) which stumulates its development into a tumor cell. Not all chemicals are initiators and not all chemicals are promoters. Dr. Spangler indicated that in their study dimethylbenzanthracene (DMBA) was a positive control for initiation and the positive control for promotion was 12-O-tetradecanoylphorbol-13-acetate (TPA) from croton oil.

Dr. Spangler further explained that initiation need occur only once, but that the promotional event apparently usually must occur many times. Furthermore, it is only effective if it occurs after the initiating event. Dr. Spangler went on to explain that some agents, complete carcinogens, appear to be able to both initiate and promote cancer development.

The cellular stages of the subsequent development of an initiated tumor cell have been called, in developmental order, latent tumor cell, quiescent tumor cell and tumor cell. In the study of Drs. Spangler and Ward, the tumor of interest was a skin papilloma, a benign tumor arising from epithelial tissue and growing on a stalk. It may progress to a carcinoma.

Dr. Spangler also discussed the modifications to this two-stage theory that have been proposed for the development of skin carcinomas. He indicated that at present it is not known whether (1) the carcinomatous cell develops from a papillomatous cell which in turn has developed from a normal cell or (2) the carcinomatous cell develops slowly and directly from the normal cell following initiation and promotion. Dr. Spangler indicated that carcinoma development took at least 40 weeks in their model.

After these introductory remarks, Dr. Spangler then presented the details of his and Dr. Ward's study. They are shown in Table 1. In Column 2 of Table 1, the first letter indicates the initiator and the second and third letters indicate the promoter for that group.

In their studies, Drs. Spangler and Ward used 3.7% solutions of formaldehyde in acetone. As shown in Table 1, they used DMBA as a positive control for initiating activity. It was applied once at a dose of 20μg per mouse. The positive control promoting agent, TPA, was applied at a dose of 1.25μg per mouse per application. There were 30 female SENCAR mice in each group.

In discussing the results of his and Dr. Ward's studies, Dr. Spangler stressed that he was reporting only papilloma incidence since the study had been in progress for only 48 of the scheduled 78 weeks. No prediction of carcinoma incidence was possible and the data were to be considered interim data only.

Dr. Spangler reported that no papillomas had developed in either of two negative control groups (Fig. 1: Groups 4A2 and 4D) or the group in which formaldehyde was being evaluated as a complete carcinogen (Table 1, Group 1). Thus, there was not yet any indication that formaldehyde was a complete carcinogen in the SENCAR mouse. At the same time, there was an average of 16.6 papillomas per mouse in the positive control group initiated with DMBA and promoted with TPA (Table 1, Group 4A1 and Fig. 1). The regression seen after Week 42 in the positive control group (Fig. 1) may be due to either or both of two factors, according to Dr. Spangler: (1) the papillomas may be coalescing or (2) they may be regressing. If carcinomas form later, the papillomas also may decrease in numbers because the carcinomas may obliterate them as the carcinomas spread over the skin surface.

Dr. Spangler then pointed out that Figure 2 appeared to suggest that formaldehyde may be a weak initiator. However, he felt that this interpretation of the data was contradicted by the data presented in Figure 4. He pointed out that the papilloma incidence and pattern of appearance did not appear to be different whether an acetone solution of formaldehyde was used as an initiator and an acetone solution of TPA as a promoter (Fig. 2) or whether acetone alone was used as an initiator and an acetone solution of TPA was used as a promoter (Fig. 4). Thus, he did not feel that the totality of the data indicated that formaldehyde was even a weak initiator.

Dr. Spangler then discussed the results for Group 3 where DMBA was used as an initiator and formaldehyde was used as a promoter. A slight increase in the number of papillomas was found (Fig. 3). A similar picture was seen with Group 4E, where acetone was used as an initiator and formaldehyde as a promoter (not shown in any figure). The time course of papilloma development was similar to that seen in the experiments where TPA was used as a promoter (Groups 2, 4A1, and 4C) but the incidence was much less.

To explain this observation, Dr. Spangler reviewed the theory of Yuspa and Hennings that the skin of the SENCAR mouse contains a few cells that are already constitutively or otherwise initiated. Promoters act on these cells and cause their development through the stages already discussed. He also pointed out that the results could be interpreted to indicate that TPA also had very weak initiating activity and formaldehyde had even weaker initiating activity. However, he felt that the present study with formaldehyde and the history of TPA both contradicted this interpretation. He felt the theory of Yuspa and Hennings, which invoked the concept of a few constitutively induced skin cells, was a more viable explanation.

Dr. Spangler concluded his remarks by indicating that he interpreted his and Dr. Ward's study to indicate formaldehyde probably is not a complete carcinogen and probably is not an initiating agent. He felt there was a slight possibility that formaldehyde may be a very, very weak promoting agent.

DISCUSSION

DR. MORGAN: (Kevin Morgan, CIIT.) Formaldehyde is extremely water-soluble, and compounds that are very water-soluble are generally poorly lipid soluble. The skin has a funny surface, and it protects you from the rain; thus, if formaldehyde is deposited on the surface of the skin, it might not actually penetrate through to the sensitive cell population underneath.

DR. SPANGLER: We used acetone for the reason that I stated at the outset. We really did not take skin penetration into consideration. Possibly we should have. Also, we selected the test concentration because, again, that is a concentration that humans are typically or potentially exposed to. We originally wanted to use an aqueous solution, but because the other agents were dissolved in acetone and because of the roll-off phenomenon that I mentioned, we decided to use acetone. But that is a good point, and we did not take that into full consideration.

DR. MORGAN: I think a negative result doesn't tell us very much, if you cannot show that it gets through the surface.

DR. BERNSTEIN: (Martin Bernstein, Ciba-Geigy.) Two questions. One, you made a statement earlier that initiation leads to a mutational event which does not lead to a tumor. I find that very interesting, because a lot of people are saying that initiation, which is a mutational event, if it is left alone, isn't going to go to a tumor.

DR. SPANGLER: Right. I was just addressing what is called the classical two-stage theory. That is one of the tenets of that theory -- you have to have both initiation and promotion. You have to have the initiation which is a mutagenic event something happening to the DNA; but you also have to have the promotional event. Initiation alone will not induce tumors; apparently in the skin system, at least. Nothing will happen unless you also promote with TPA or another promoting agent.

DR. BERNSTEIN: Then, theoretically, you can have a genotoxic event without having a subsequent carcinogenic event.

DR. SPANGLER: I would presume so.

DR. BERNSTEIN: Second question: How does the skin promotion system relate to inhalation exposure, since we are dealing with formalin solutions rather than formaldehyde gas?

DR. SPANGLER: We were interested because of the fact of human skin exposure, particularly to the hands; people working in histopathology, etc. I failed to mention that we will also look at all the internal organs, including nasal cavities at terminal sacrifice, which will occur after 10 months. All internal organs will be evaluated histopathologically as well.

In the first week of promotion, when the animals were housed initially in cages that had filter tops, we noticed that the animals were running around acting very strangely in the cage or they would end up sitting in the corner. They were actually holding their breath, and I don't blame them. We believe this was because of the fact that the formaldehyde was not escaping rapidly from the cages. We quickly got rid of the filter tops and went to wire tops. From that point of view they were exposed to the vapors in the first week of promotion. For the rest of the study, the animals, once promoted, were left under the hood. Formaldehyde evaporates very rapidly in the hood and the animal is put back on the shelf. We know there is limited exposure through the inhalation route, because when they are brought back out in the room, there is no odor whatsoever in the room from the formaldehyde.

Of course, the human can smell perhaps, 0.5 to 1 part per million of formaldehyde in the air.

CHAPTER 6

SKIN INITIATION-PROMOTION STUDY WITH FORMALDEHYDE IN CD-1 MICE

N. D. Krivanek, N. C. Chromey and J. W. McAlack

E.I. du Pont de Nemours & Company, Inc.
Wilmington, Delaware

Formaldehyde (HCHO) in a 50:50 acetone:water vehicle was tested for its ability to initiate and/or promote tumorigenesis in CD-1 female mice. For evaluation of formaldehyde initiator activity, a single dose of five mg HCHO in 50 µl vehicle was used. Formaldehyde promoter doses were 1.0, 0.5 and 0.1 mg in 100 µl vehicle. Appropriate positive and negative controls were included. Treatment groups consisted of 30 animals each.

A single initiation treatment was followed two weeks later by promoter treatment three times weekly for 180 days. Animals were observed another 180 days before they were sacrificed. There were no significant differences in mortality between control and treated groups. Mortality was low for all groups.

Skin nodules were tallied weekly. Nodules persisting for less than 30 days were not counted. Incidence of skin nodules at end-of-study for the following initiator/promoter combinations were: acetone/phorbol myristate acetate (TPA) -- 3/29; benz(a) pyrene (BaP)/TPA -- 28/29; HCHO/TPA -- 5/29; HCHO/acetone -- 0/30; HCHO/HCHO -- 0/30; BaP/1.0 mg HCHO -- 1/30; BaP/0.5 mg HCHO -- 2/30; BaP/0.1 mg HCHO -- 7/30; and BaP/acetone -- 3/29.

Skin sections from all mice are being evaluated histopathologically.

INTRODUCTION

Formaldehyde has produced tumors in the nasal cavity of rats and mice exposed to high concentrations of formaldehyde vapors in long-term inhalation experiments (Swenberg et al.[1]) The actual mechanism of formaldehyde's carcinogenic action has not been elucidated, but research by scientists at the Chemical Industry Institute of Toxicology has produced evidence that the development of tumors from formaldehyde exposure may involve a multistep process which includes cell damage and cell proliferation.

The existence of a multi-stage process for cell transformation in cultured cells which may mimic the *in vivo* process has been studied with formaldehyde. Studies by Boreiko et al.[2,3] with the C3H10T1/2 cell culture assay have shown that formaldehyde was both a weak initiator and promoter of cell transformation. In whole animals an assay system has been developed which can determine the ability of a chemical to initiate or promote tumors. It was the objective of this study to use a modified version of the mouse skin-painting assay described by Van Duuren et al.[4] to determine if formaldehyde will initiate or promote skin tumors.

This study is divided into two parts. The first part is a repeated exposure skin irritation pretest and the second an initiation-promotion study with formaldehyde solutions and appropriate controls. The purpose of the pretest was to determine irritating and nonirritating doses of formaldehyde. To minimize skin damage which could obscure test results, no more than slightly irritating concentrations were used in the initiation-promotion study.

METHODS

Animals

Female CD-1 mice (Charles River Breeding Laboratory) were used for this study. Females were used because we have had good experience with them in previous studies at our laboratory. The mice were individually housed in suspended stainless steel wire mesh cages during the study. The mice received water and Certified Purina Laboratory Chow #5002 *ad libitum*. The mice were 22 days old when received and during a three-week acclimatization period they were observed for weight gain and gross signs of disease or injury. At the end of this period, the mice were divided by computerized, stratified randomization into nine groups of 30 mice each such that the mean body weights of each group were statistically equal. After assignment to treatment groups, all mice were ear-clipped and toe-clipped for permanent individual identification.

Test Materials

Benzo(a)pyrene (BaP) 99% pure (J. T. Baker) was used as the positive initiator agent. Phorbol myristate acetate (12-O-tetra-decanoylphorbol-13-acetate; TPA) 99% pure (Consolidated Midland Corp.) was used as the positive promoter. Acetone 99.7% pure (J. T. Baker) was the solvent. Paraformaldehyde powder 96.8% pure (Celanese Chemical Company) was used to prepare formaldehyde solutions.

The formaldehyde stock solutions were made by adding paraformaldehyde powder to distilled water in a sealed Reacti-Flask[R]. These solutions were stirred and heated at 75°C in a water bath for 4 hours. A stock solution of 20% formaldehyde was prepared. This 20% formaldehyde solution was diluted with acetone to produce a 10% formaldehyde solution in a 50:50 acetone:water mixture. Analyses of this 10% solution by nuclear magnetic spectroscopy over a period of 6 days did not show any indication of a formaldehyde-acetone condensation product. The other formaldehyde solutions were prepared from a 2% formaldehyde aqueous stock solution which was prepared by the same procedure used for the 20% stock solution. Dilutions to lower concentrations were made with appropriate amounts of distilled water and/or acetone. Formaldehyde test solutions were stored under refrigeration and prepared fresh weekly. Standard solutions were analyzed according to the modified NIOSH method P&CAM 125.[5]

Skin Irritation Test

Formaldehyde solutions were mixed with acetone in a 50:50 acetone:water ratio to facilitate spreading the test material on the skin. Animals were shaved with an electric animal clipper as needed and the test materials were applied to an approximate 1 inch square area on the shaved back of the mice. Animals were treated daily, except for weekends, for two weeks with the 10.0 (10%), 5.0 (5.0%), 2.0 (2.0%) or 1.0 (1.0%) mg per 100μl acetone:water doses. The 0.5 (0.5%) and 0.1 mg (0.1%) per 100μl acetone:water doses were applied to the mice for three weeks. Skin responses were observed daily, except for weekends.

Initiation-Promotion

The study design is shown in Fig. 1. The dosing regimen is designated by "initiator dose-promoter dose." Initiation treatment consisted of a single application of the indicated test material. Promotion treatments were started two weeks after the initiation treatment and consisted of application of promoter three times per week for 26 weeks. The volume of material applied for initiation treatment was 50 μl and for promotion treatment 100 μl. The solvent system for formaldehyde was 50:50/ acetone:water. BaP and TPA were dissolved in acetone.

Three control groups were used:

1. Initiator control - 50 μl acetone/2.5 μg TPA
2. Promoter control - 150 μg BaP/acetone
3. Positive control - 150 μg BaP/2.5 μg TPA

Formaldehyde was tested as an initiator, a promoter, and a complete tumorigen. The initiator potential of formaldehyde was tested in the 5 mg (10%) formaldehyde/2.5 μg TPA group and the control for this group was 50 μl acetone/2.5 μg TPA. The promoter potential of formaldehyde was tested at three dose levels: 150 μg BaP/1.0 mg formaldehyde, 150 μg BaP/0.5 mg formaldehyde and 150 μg BaP/0.1 mg formaldehyde. The control group for the formaldehyde promotion groups was the 150 μg BaP/acetone group. The doses used to test formaldehyde as both an initiator and promoter were 5 mg formaldehyde/1.0 mg

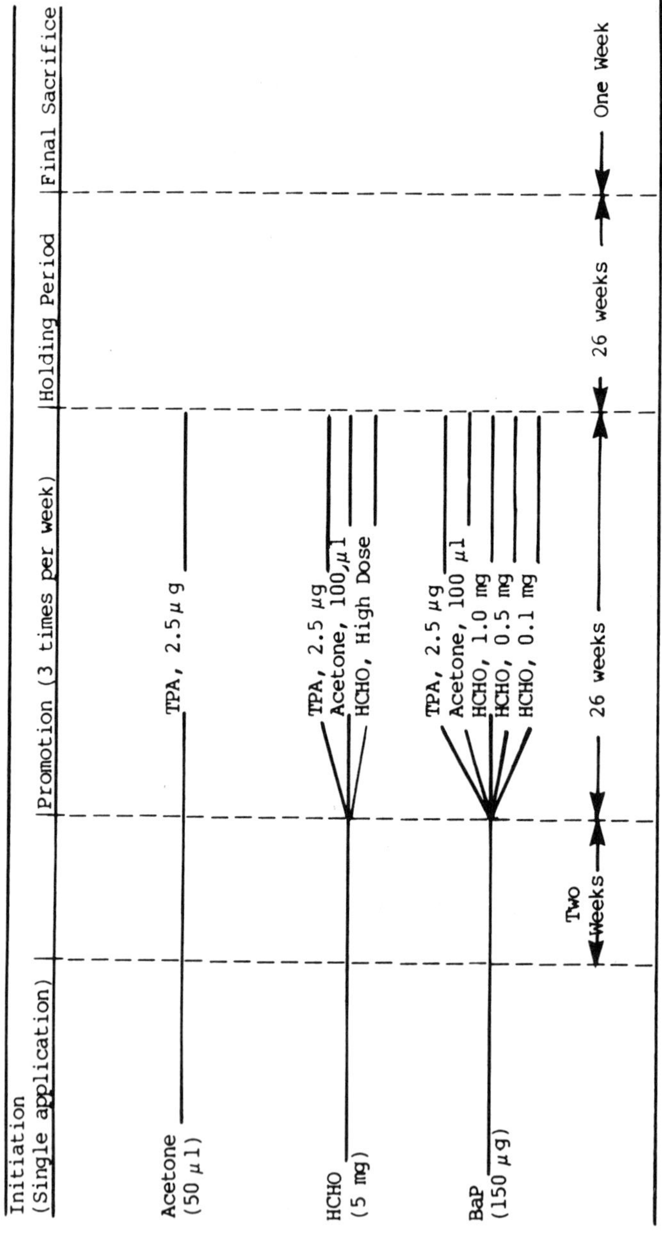

Fig. 1. Flow chart for this initiation-promotion study.

formaldehyde. A control for this group was the 5 mg formaldehyde/acetone group. After 26 weeks of promotion treatments the mice were held for an additional 26-week recovery period.

In-life observations made during the study are described below. Body weight was measured biweekly during promotion and monthly during the holding period. Skin test sites were observed daily during promotion and biweekly during the holding period. Skin nodules were charted monthly. A nodule had to persist for at least 30 days to be included in the final tally.
Mortality was recorded as it occurred and moribund animals were sacrificed *in extremis*.

RESULTS

Skin Irritation Pre-Test

The 10 mg doses of formaldehyde produced moderate skin irritation after 2-4 applications characterized by drying and cracking of the skin (fissuring), sloughing, and papules. The 5.0 and 2.0 mg doses produced mild to moderate skin irritation after 4-5 treatments. The 1 mg dose produced mild irritation which began during the second week of treatment. At 0.5 mg, slight irritation occurred, but recovery to normal was observed after the weekend rest period. At 0.1 mg, no skin irritation was observed even after three weeks of treatment. No mortality or any other signs of toxicity were observed. Doses of 1.0, 0.5, and 0.1 mg were selected as the three dose levels of formaldehyde for promotion treatments based upon their skin irritation potential.

Initiation-Promotion Study

The body weight curves showed no difference between groups during the course of the study. Fig. 2 shows these growth curves for four representative groups: acetone/TPA, BaP/TPA, 5 mg formaldehyde/TPA and 5 mg formaldehyde/1 mg formaldehyde. These curves are typical for all groups and no significant differences were found. Mortality was low in all but the positive control group ranging from 2-5/group. Eight positive control mice died or were sacrificed *in extremis*, most due to large ulcerated tumors.

The incidence of skin nodules is shown in Table 1. The acetone/TPA groups had 3/27 mice with nodules compared to 5/28 mice in the formaldeyde/TPA group. Comparison of these groups for the initiator potential of formaldehyde revealed no significant difference. In the positive control group, BaP/TPA, 29/29 mice had at least one skin nodule. The incidence of nodules for the BaP/acetone group was 3/27; for BaP/0.1 mg formaldehyde, 7/27; for BaP/0.5 mg formaldehyde, 2/28; and for BaP/1.0 mg formaldehyde, 1/25. When analyzed by Fisher's exact test there is no statistical difference between these groups. The biological significance of the inverse dose response relationship observed among the formaldehyde promoted groups is unclear. There were no mice with nodules in the formaldehyde/ formaldehyde group which tested for combined initiation/promotion potential. The control group formaldehyde/acetone also had no mice with nodules.

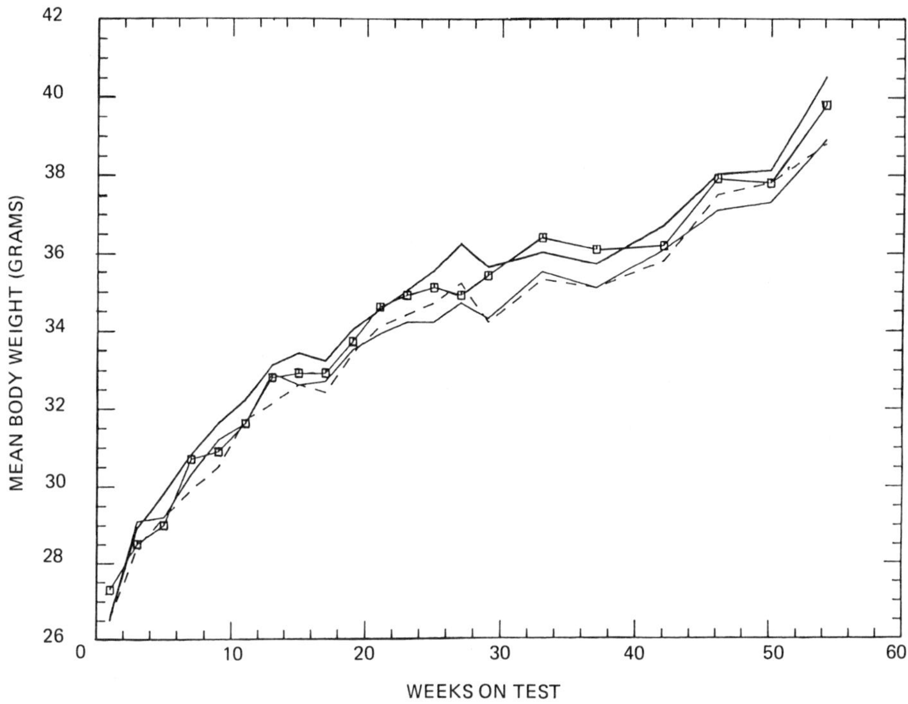

Fig. 2. Growth curves for the mice in this test. ——, acetone/TPA treated mice; ▬▬, BaP treated mice; -----, 5 mg formaldehyde/TPA treated mice; -□-□-, 5 mg formaldehyde/1 mg formaldehyde treated mice.

TABLE 1

Incidence of Skin Nodules at Test Site in CD-1 Female Mice

Initiation-Promotion	Mice With Modules/Mice at Risk	No. of Nodules	Avg. Nodules/Nodule Bearing Mouse	Days to First Nodule	Avg. Days to First Nodule for Group
Acetone/TPA	3/27	7	2.3	147	355
BaP/TPA	29/29	149	5.1	65	110
HCHO/TPA	5/28	6	1.2	90	341
HCHO/Acetone	0/28	0	0	---	378
HCHO/HCHO	0/28	0	0	---	378
BaP/HCHO-High	1/25	1	1	171	370
BaP/HCHO-Mid	2/28	2	1	167	368
BaP/HCHO-Low	7/27	9	1.3	146	338
BaP/Acetone	3/27	3	1	163	361

Finally, Table 1 summarizes time-to-first-nodule and the average time to onset of the first nodule for each group. This shows that the positive control, BaP/TPA, has an earlier time-to-nodule onset than the other groups.

Skin Histopathology

During the week of study termination all skin nodules present at that time were excised, stained, and examined histopathologically. The results of this examination are summarized in Table 2. In the acetone/TPA, BaP/TPA, formaldehyde/TPA, BaP/1.0 mg formaldehyde and BaP/0.1 mg formaldehyde group the number of nodules does not equal the combined total tumor incidences. These discrepancies resulted from the following reasons. For one nodule in both the acetone/TPA and the BaP/0.1 mg formaldehyde and two in both BaP/TPA and formaldehyde/TPA group, nodules were less than 3 mm in diameter and were lost during tissue processing. For one nodule in the BaP/1.0 mg formaldehyde and two in the BaP/TPA group, the nodules had persisted for more than 30 days and therefore were counted in these groups, but they regressed by the time of necropsy. For two nodules in BaP/TPA group, the nodule had undergone extensive necropsy which precluded precise histopathologic diagnosis. On the basis of the gross findings these nodules were probably malignant tumors.

A major finding of the study was that only the positive control group developed malignant tumors--squamous cell carcinomas. The benign tumors were either keratoacanthomas or

TABLE 2

Incidence of Histologically Confirmed
Skin Tumors at Test Site in CD-1 Female Mice

Group Number: Initiator-Promoter	Number of Mice with Nodules	Number of Mice with Benign Tumors*	Number of Mice with Malignant Tumors
1. : Acetone/TPA (29)**	3	2	0
2. : BaP/TPA (29)	28	15	9
3. : HCHO/TPA (30)	5	3	0
4. : HCHO/Acetone (30)	0	0	0
5. : HCHO/HCHO (29)	0	0	0
6. : BaP/HCHO-High (29)	1	0	0
7. : BaP/HCHO-Mid (30)	2	2	0
8. : BaP/HCHO-Low (30)	7	6	0
9. : BaP/Acetone (30)	3	3	0

* Keratoacanthoma or squamous papilloma.
**Number of mice examined.

squamous papillomas; their incidence was equally distributed for a given group. There were no statistically significant differences found when test groups were compared with appropriate controls using Fisher's Exact Test.

DISCUSSION

The results of this study show that minimally irritating solutions of formaldehyde do not initiate nor promote skin tumors nor does formaldehyde act as a complete tumorigen when evaluated in this mouse skin painting test.

This study also confirms the ability of BaP to act as a tumor initiator and TPA as a tumor promoter in this assay. Malignant skin tumors were found only in the positive control group, and not in any of the other test groups.

ACKNOWLEDGMENTS

Thanks to Dr. R. W. Hartgrove for help in study design and consultation. This study was sponsored by the Formaldehyde Institute.

REFERENCES

1. J. A. Swenberg, W. D. Kerns, R. I. Mitchell, E. J. Gralla, and J. L. Pavkov, Induction of squamous cell carcinomas of the rat nasal cavity by inhalation exposure to formaldehyde vapor. Cancer Res. 40, 3398 (1980).

2. C. J. Boreiko, D. B. Couch, and J. A. Swenberg. "Mutagenic and Carcinogenic Effects of Formaldehyde in Genetic Effects of Airborne Agents," R. Tice, D. L. Costa, and K. Schaich, eds., Plenum Press, New York, 1982, p. 353.

3. C. J. Boreiko, D. J. Abernethy, and J. H. Frazelle. Promotion of C3H/10T1/2 Cell transformation by formaldehyde, Proc. Environ. Mutagen Soc., 13, 120 (1982).

4. B. L. Van Buuren, A. Sivak, A. Segal, I. Seidmen, and C. Katz, "Dose-response studies with a pure tumor-promoting agent." Cancer Res. 33, 2166 (1973).

5. NIOSH, NIOSH Manual of Analytical Methods, Second Edition, Volume 1, P&CAM No. 125, April, 1977, DHEW (NIOSH), Washington, D.C., Publication No 77-157A.

DISCUSSION

DR. BERNSTEIN: (Martin Bernstein, Ciba-Geigy.) I have somewhat the same questions I asked Dr. Spangler. First, the relationship of this type of study to inhalation exposure and your feeling on that? Second, how frequently did you analyze the solutions that you prepared; how frequently did you prepare them; and what is their stability?

DR. KRIVANEK: The first question -- as far as a direct connection between the initiation-promotion on the skin and the tumors seen in the CIIT study -- I don't have specific comments on that. All I can say is that this is another test system, and it is more of a model system rather than a whole animal experiment, where one would like to test formaldehyde on the skin for a full lifetime of the animal at a variety of doses. That experiment has not yet been done.

As far as the formaldehyde solutions, these were prepared from the para-formaldehyde powder. A stock solution was made and then that was diluted. I do not have the frequency of analysis here with me. The concentrations were low, and such low concentrations are known to be stable. In addition, they were refrigerated and the solutions were always clear. So, we felt that we weren't losing any formaldehyde in the study.

DR. BERNSTEIN: But it wasn't analytically confirmed?

DR. KRIVANEK: I don't know if it was or not. Perhaps Mr. McAlack, who carried out the study, is in the audience and will know the answer.

MR. MC ALACK: (John Mc Alack, Haskell Laboratory, DuPont.) No, we didn't. We had an analytical group do initial analyses, and we prepared them by the same method weekly.

CHAPTER 7

MUTAGENIC EFFECTS OF FORMALDEHYDE IN BACTERIAL AND HUMAN CELLS

V.S. Goldmacher, P. Temcharoen* and W.G. Thilly

Massachusetts Institute of Technology
Cambridge, Massachusetts

Formaldehyde is known to be mutagenic for bacteria and for some other organisms. Recently we reported the first data on formaldehyde mutagenicity for human cells. We found formaldehyde to be mutagenic for diploid human lymphoblasts in culture. The minimum concentration required to induce appearance of a significant number of trifluorothymidine-resistant mutants were 130μM or 4 ppm by weight (2 hr. exposure at 37°C). Our results suggest that formaldehyde may be a mutagenic hazard for humans.

INTRODUCTION

Formaldehyde is widely used in industry and medicine [1,2,3], has been found to be an urban and industrial air pollutant [2,4], and is found in tobacco smoke[5,6,7], in products of the thermal degradation of polymeric materials[8], in automobile and diesel exhaust fumes[9,10], and in incinerator effluents[11].

In recent years, formaldehyde has attracted extensive attention due to its suspected role as a mutagenic and carcinogenic risk. This suspicion is based on several facts.

*Current Affiliation: Department of Pathobiology, Faculty of Science, Mahidol University, Bangkok, Thailand.

Exposure of rats and mice to formaldehyde vapor in laboratory studies causes carcinogenic effects [2,12]. The compound has been found to be mutagenic for some bacteria, insects, flowering plants, and fungi[13]. Formaldehyde induced sister-chromatid exchanges in Chinese hamster ovary cells and in human lymphocytes[14] and also induced unscheduled DNA synthesis in HeLa cells[15]. It is not yet clear, however, whether formaldehyde presents a hazard for humans, since other investigations have produced conflicting results. For example, testing of formaldehyde mutagenicity in mice, performed by Epstein and Shafner[16], yielded negative results. Also, and more importantly, there has been no direct evidence that formaldehyde is mutagenic or carcinogenic in humans or in cells of human origin.

In order to provide a basis for further studies on the interaction of formaldehyde with human cells, we measured mutagenicity and toxicity of formaldehyde in a human lymphoblastoid line, using forward mutation to trifluorothymidine resistance (F_3TdR^R). For comparison, mutagenicity of formaldehyde was tested in _Salmonella typhimurium_, using forward mutation leading to 8-azaguanine (8-AG) resistance.

METHODS

Chemicals

Formaldehyde (analytical grade) 37% (12.3 M) aqueous solution with 10%-15% (approximately 3-5 M) of methyl alcohol (stabilizer) was purchased from Anachemia LTD (Montreal,

Canada). Methyl alcohol (analytical grade) was obtained from Mallinckrodt, Inc. (St. Louis, MO). Aroclor (Registered trade mark of Monsanto Corp. for polychlorinated biphenyl)-induced postmitochondrial supernatant (PMS) was purchased from Litton Bionetics (Kensington, MD). N-methyl-N'-nitro-N-nitrosoguanidine (MNNG), glucose-6-phosphate, NADP, glucose-6-phosphate dehydrogenase, cytidine, hypoxanthine, thymidine, aminopterin, 4-nitroquinoline-N-oxide (4NQO), trifluorothymidine (F_3TdR), and 8-azaguanine (8-AG) were purchased from Sigma Chemical Co. (St. Louis, MO). All inorganic salts were analytical grade.

Media for Lymphoblast Cultures

Cells were grown and plated in RPMI 1640 medium (Flow Laboratories, McLean, VA) supplemented with 5% horse serum (Flow Labs). The horse serum was heat-treated at 56°C for 2 hrs. before use in order to eliminate a factor which degraded F_3TdR. CHAT medium consisted of RPMI 1640 serum supplemented medium with 10^{-5} M cytidine, 2 x 10^{-4} M hypoxanthine, 2 x 10^{-7} M aminopterin and 1.75 x 10^{-5} M thymidine. THC medium was CHAT without aminopterin. Both solutions were stored frozen as 100x aliquots.

Lymphoblast Cell Culture Maintenance

The lymphoblast line TK6, a clonal derivative of the WI-L2 line, was isolated from a male donor with hereditary spherocytosis[17]. The TK6 line is heterozygous at the thymidine kinase locus $tk^{+/-}$[18]. We measured frequency of the forward $tk^{+/-} \rightarrow tk^{-/-}$ mutation. Mutants which are $tk^{-/-}$ lack active thymidine kinase (TK); they are unable to utilize the thymidine analogue F_3TdR and are thus resistant to the toxic effects of the compound. The toxic and mutagenic responses of the TK6 cell line to different chemicals have been extensively studied in our laboratory[18-20].

The cells were maintained in exponential growth (doubling time 17 hr.) in stationary culture. Cultures were passaged by daily dilution to 3 x 10^5 cells/ml. Cell counts were made on a Coulter counter (Coulter Electronics, Hialeah, FL).

In order to observe colony formation from single cells, cultures were plated in microtiter plates (Linbro Division of Flow Laboratories, Hamden, CO). The lymphoblasts formed colonies of about 1-2 x 10^5 cells.

Mutation Assay

In order to reduce background mutant fraction of spontaneous mutants, cultures were treated in CHAT medium 4-7 days before each mutation experiment. Aminopterin inhibits the single carbon transfer by folic acid reductase necessary for the *de novo* synthesis of purines and thymidine monophosphate. Cells containing thymidine kinase (*tk*$^{+/-}$ cells) grow in the presence of aminopterin using the exogenous DNA precursors, but pre-existing *tk*$^{-/-}$ mutants cannot. Following 48 hr. CHAT treatment, cells were centrifuged and placed into TCH medium in order to maintain exponential growth[21]. Cells were diluted with regular medium on the ensuing days, gradually diluting away the THC.

After this treatment, 2-12 x 10^7 cells (4 x 10^5 cells/ml) were exposed to different concentrations (0-150μM) of formaldehyde for 2 hrs. at 37°C. Exposure was terminated by centrifuging the cells and resuspending them in fresh medium. The cells were then maintained by daily dilutions to 3 x 10^5 cells/ml for 4-6 days, a sufficient time for phenotypic expression of F_3TdR resistance [19].

MNNG (0.1μM x 45 min), shown to be a potent mutagen for human lymphoblasts, was used as a positive control. MNNG induced mutations with frequency consistent with previous studies with TK6. Negative control consisted of the incubation mixture without formaldehyde. Cells were also exposed to 70μM methanol as a negative control for the formaldehyde solution.

Two methods were used for determining cell survival after treatment with formaldehyde: 1) estimation of clone forming ability of treated cells in comparison with untreated cells, and 2) estimation of the number of survivors in the bulk culture from growth curve extrapolation. Immediately after treatment with formaldehyde, 500 or more treated and untreated cells were plated into microtiter wells at a cell density of 3 cells/well (untreated cells) or 3-100 cells/well (treated cells) in 0.2 ml of the medium. For mutation experiments, in order to allow phenotypic expression and to allow recovery from the toxic effects of formaldehyde, the cells were grown 406 days following treatment with formaldehyde.

Cells were grown in the presence (3-4 plates per point) and absence (3 plates per point) of the selective agent F_3TdR (2μg/ml). Three cells were plated per well in 0.2 ml medium in nonselective conditions, and 4 x 10^4 cells/well in selective conditions. The plates were incubated at 37°C in a humidified incubator with 5% CO_2 for 13 days until colonies formed, after which colonies were counted for mutant fraction determination.

Mutant frequencies were calculated as follows. According to the Poisson distribution, the average number of colony-forming units per well (λ) is λ = -ln(fraction of observed negative wells). The clone-forming ability of a culture (plating efficiency, PE) is:

$$PE = \frac{\lambda}{\text{average number of cells plated per well}}.$$

PE of untreated TK6 cells ranged from 35%-75%. The procedure for calculating the mutant fractions and their 95% confidence intervals has been presented elsewhere[21].

The mutant fraction (MF) was calculated as:

$$MF = \frac{\ln(x_s/n_s)}{\ln(x_o/n_o)} \times \text{dilution factor, where}$$

x_o - Number of empty wells in non-selective conditions

n_o - Total number of wells scored in non-selective conditions

x_s - Number of empty wells in selective conditions

n_s - Total number of wells scored in selective conditions

The 95% confidence intervals are:

$$MF \times \exp\left[\pm Z\, a_{/2} \sqrt{\frac{1-(x_s/n_s)}{x_s \ln(x_s/n_s)^2} + \frac{1-(x_o/n_o)}{x_o \ln(x_o/n_o)^2}}\right]$$

Bacterial Mutation Assay

This assay is based on the ability of mutants of *Salmonella typhimurium*, lacking the functions of either membrane translocation or phosphoribosyl-transferase, to form colonies in agar containing 8-AG (50 μg/ml). Wild type cells cannot form colonies in the presence of 8-AG. The Salmonella typhimurium strain used was TM677, a his+ revertant of Ames' strain TA1535, which carries plasmid pKM 101. The development and initial use of this assay are described elsewhere [18,23].

In each assay, a 1 ml frozen aliquot, containing 5 x 10^8 viable cells, was quickly thawed and immediately diluted with 50 ml of Minimal E medium (0.2 mg/ml $MgSO_4 \cdot 7H_2O$, 2 mg/ml citric acid • H_2O, 10 mg/ml K_2HPO_4, 3.5 mg/ml $NaNH_4HPO_4 \cdot H_2O$, 2% dextrose, 0.01 mg/ml biotin, pH 7.0) at 37°C. Doubling time was approximately 40 minutes. In metabolic activation experiments, 0.10 ml (10% of final volume) of Aroclor-induced postmitochondrial supernatant (PMS) (containing 24.5 mg/ml protein), glucose-6-phosphate (1 mg/ml), NADP (1 mg/ml), $MgCl_2$ (670 mg/ml), and glucose-6-phosphate dehydrogenase (0.4 unit/ml), were added into the cultures for the drug metabolizing system. 1.0 ml bacterial cultures were exposed to the various concentrations of formaldehyde in 15 ml centrifuge tubes and incubated for 15-120 min at 37°C in a rotor wheel. Following incubations, the reaction mixtures were diluted fivefold by the addition of 4 ml room-temperature phosphate-buffered saline to each centrifuge tube. The appropriate dilutions of the reaction mixtures were then plated under selective conditions (50 μg/ml 8-AG) and nonselective conditions. Colonies were counted after 2 days of incubation at 37°C. Positive control consisted of exposure to 100 ng/ml 4NQO for 15-120 min. Negative control consisted of exposure to the incubation mixture without formaldehyde but with the addition of ethanol to correspond to the amount of methanol added in the formaldehyde solution to the treated cultures.

The mutant fraction was calculated by dividing the number of colonies observed under selective conditions by the number of colonies observed under permissive conditions and multiplying by appropriate dilution factors[18,23]. The data represent the average of at least two independently treated cultures.

RESULTS AND DISCUSSION

Results presented in Table 1 show that 150μM formaldehyde induced a significant number of F_3TdR resistant mutants in the human lymphoblast cells. The minimum detectable concentration of formaldehyde which induced mutation to F_3TdR^R was about 130μM, or 4 ppm (Fig. 1).

Toxicity of formaldehyde to TK6 cells, estimated both by clone-forming ability of treated cells compared to untreated cells and by growth curve extrapolation, is illustrated in Fig. 1 (top). It is evident that the apparent surviving fractions obtained by the two methods are different. However, the differences probably did not affect our estimates of the mutant fractions, since the cells were allowed to recover from these toxic effects of the mutagen treatment before plating for mutant fraction determination.

The possibilty that the methanol additive in formaldehyde was the cause of the observed mutation for the TK6 cells was tested by treating the cells with the approximate concentration of

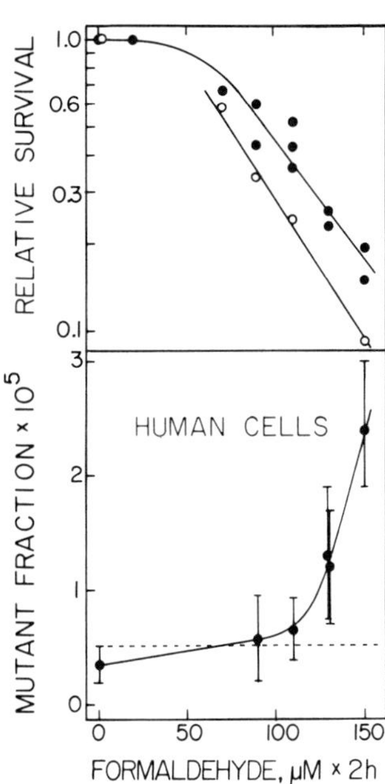

Fig. 1. Mutagenic (bottom) and toxic (top) activities of formaldehyde to human cells. Vertical bars are 95% confidence limits, based on two independently treated cultures. The dotted line represents the 95% upper limit on the observations on untreated controls. Relative survival of the cells after formaldehyde treatment was estimated by measurements of clone-forming ability (○) and from the growth curves of the treated cell cultures (●).

TABLE 1

Spontaneous and Induced F_3TdR Resistant Lymphoblast Mutant Fractions

Treatment	Mutant Fraction x 10^5 Mean	LL[a]	UL[b]	Surviving Fraction	Estimated No. of Surviving Mutant Cells
150 μM Formaldehyde x 2 hr.	2.40	1.90	2.99	0.15	403
Control	0.34	0.19	0.51	1.00	286
70 μM Methanol x 2 hr.	0.28	0.14	0.42	1.00	268
0.1 μM MNNG x 45 min.	4.90	4.26	5.65	0.05	175

[a]Lower and [b]upper 95% confidence limits

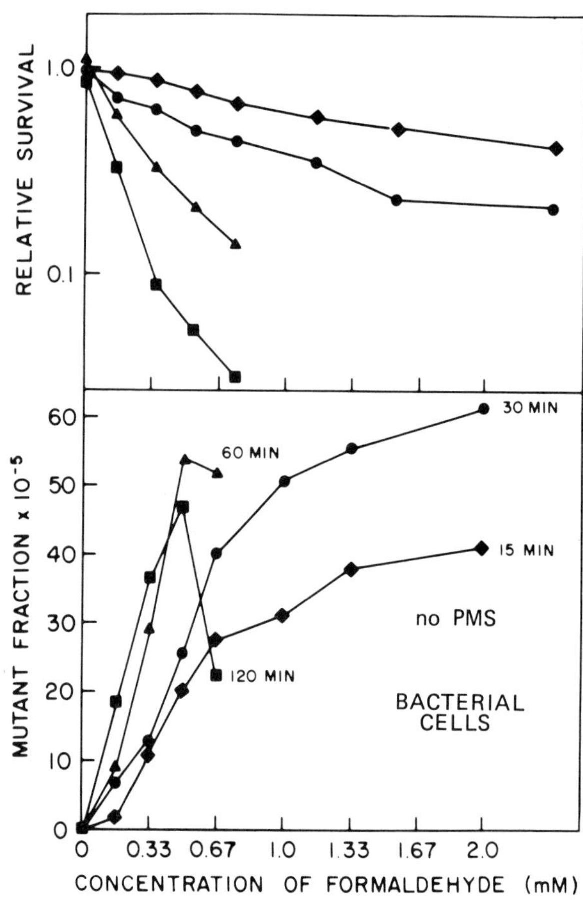

Fig. 2. Concentration and time dependent toxicity and mutagenicity of formaldehyde to S. Typhimurium in the absence of Aroclor-induced PMS, at various concentrations and exposure times: (◇) 15 min, (○) 30 min, (△) 60 min, and (□) 120 min exposure. The data for mutation fraction in each point are calculated by subtracting the background mutant fraction (approx. 10^{-4}) from the treated cultures. Duplicate bacterial cultures were independently exposed for each point. The range of variation among duplicate cultures was within ± 20% of the mean value; averages are reported here and in Fig. 3.

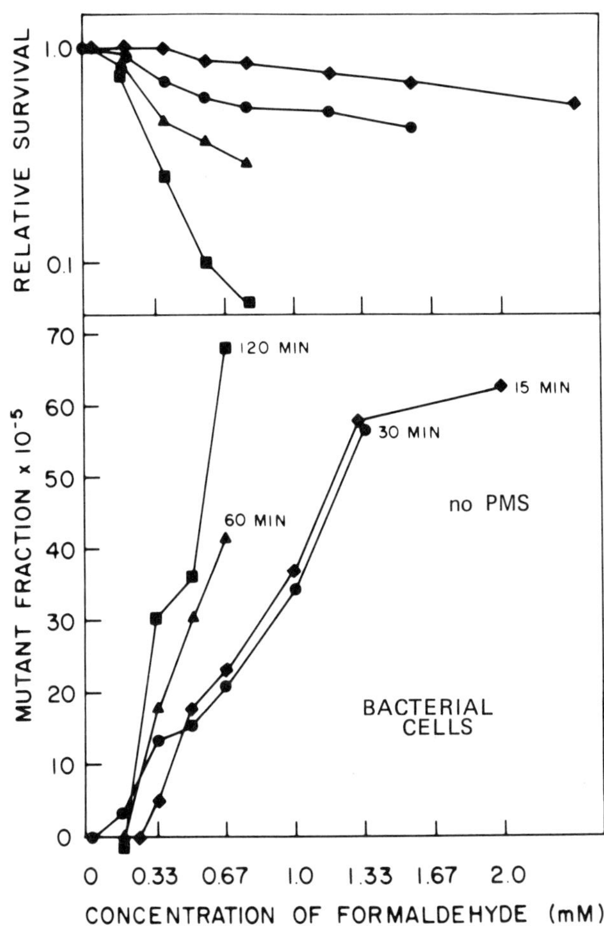

Fig. 3. Concentration and time-dependent toxicity and mutagenicity of formaldehyde to S. typhimurium was assayed in the presence of Aroclor-induced RMS at various concentrations and exposure times: (◇) 15 min, (○) 30 min, (△) 60 min, and (□) 120 min exposure. Values were calculated as for Figure 2.

methanol present in the 150μM formaldehyde solution used in this study. Exposure of the cells for 2 hr. to 70μM methanol neither killed cells nor induced detectable mutation (Table 1).

The results of the study of toxicity and mutagenicity of formaldehyde in *Salmonella typhimurium* show that the minimum concentration required to induce mutagenicity is 170μM (Fig. 2). Addition of PMS reduced both toxicity and mutagenicity of formaldehyde in the bacterial cells (Fig. 3), due probably to detoxification of formaldehyde by PMSW through oxidation[24,25] or simple Schiff-base formation with the amino groups of the PMS proteins[26].

Our results lend credence to the suspicion that formaldehyde may be carcinogenic and mutagenic for humans. It would be interesting to study the effect of multiple exposure to lower formaldehyde concentrations using this assay, since this protocol might reflect more accurately the pattern of exposure of human cells to formaldehyde *in vivo*. Our previous studies of the alkylating agents ethyl methane sulfonate and methyl nitrosourea[27] would lead us to expect a simple linear relationship between mutations and exposure at lower formaldehyde concentrations, i.e., 10 exposures to 20μM would be expected to give a mutant fraction equivalent to one exposure at 200μM.

ACKNOWLEDGMENTS

The work presented here was supported by U.S. Department of Energy Grant DE-AC02-EV267. PT was supported as a visiting scientist by IAEA Fellowship Program SC/202/THA/8133(8004).

REFERENCES

1. L. Fishbein, Environmental sources of chemical mutagens. II. Synthetic mutagens, Adv. Modern Tox. 5, 253 (1978).

2. M. Blackwell, H. Kang, A. Thomas and P. Infante, Formaldehyde: evidence of carcinogenicity, J. Amer. Ind. Hyg. Assoc., 42, A34 (1981).

3. N. I. Sax., ed, "Cancer Causing Chemicals," Van Nostrand Reinhold, New York, 1981, p. 371.

4. D. Hoffman and E. L. Wynder, in "Chemical Carcinogens," American Chemical Society, Washington, D.C. p. 324 (1976).

5. J. S. Osborne, S. Adamek and M. E. Hobbs, Some components of gas phase of cigarette smoke, Anal. Chem., 28, 211 (1956).

6. J. D. Mold and M. T. McRae, The determination of some low molecular weight aldehydes and ketones in cigarette smoke with 2, 4-dinitrophenylhydrazone, Tobacco Sci., 1, 40 (1957).

7. J. R. Newsome, V. Norman and C. H. Keith, Vapor-phase analysis of tobacco smoke, Tobacco Sci., 9, 102 (1965).

8. J. M. Stuart and D. A. Smith, Degradation of epoxide resins, J. Appl. Polym. Sci., 9, 3195 (1965).

9. D. W. Wilson, Fixation of atmospheric carbonyl compounds by sodium bisulfite, Anal. Chem., 30, 1127 (1960).

10. A. P. Altshuller, I. R. Cohen, M. E. Meyer and A. F. Wartburg, Jr., Analysis of aliphatic aldehydes in source effluents in the atmosphere, Anal. Chem. Acta, 25, 101 (1961).

11. R. L. Stenburg, R. P. Hangebrauck, D. J. V. Von Lehmden and A. H. Rose, Jr., Effects of high-volatile fuel on incinerator effluents, J. Air. Pollut. Control Assoc., 10, 114 (1961).

12. J. A. Swenberg. W. D. Kerns, R. I. Mitchell, E. J. Gralla and K. L. Pavkov, Induction of squamous cell carcinomas of the rat nasal cavity by inhalation exposure to formaldehyde vapor, Cancer Res., 40, 3398 (1980).

13. C. Auerbach, M. Moutschen-Dahmen and J. Moutschen, Genetic and cytogenetical effects of formaldehyde and related compounds, Mutat. Res., 39, 317 (1977).

14. G. Obe and H. Ristow, Mutagenic, cancerogenic and teratogenic effects of alcohol, Mutat. Res., 65, 229 (1965).

15. C. N. Martin, A. C. McDermid and R. C. Garner, Testing of known carcinogens and non-carcinogens for their ability to induce unscheduled DNA synthesis in HeLa cells, Cancer Res., 38, 2621 (1978).

16. S. S. Epstein and H. Shafner, Chemical mutagens in the human environment, Nature, 219, 385 (1968).

17. J. A. Levy, M. Virolainen and V. Defendi, Human lymphoblastoid lines from lymph node and spleen, Cancer, 22, 517, (1968).

18. T. R. Skopek, H. L. Liber, B. W. Penman and W. G. Thilly, Isolation of a human lymphoblastoid line heterozygous at the thymidine kinase locus: possibility for a rapid human cell mutation assay, Biochem. Biophys. Res. Commun., 84, 411 (1978).

19. W. G. Thilly, J. G. DeLuca, E. E. Furth, H. Hoppe IV, D. A. Kaden, J. J. Krolewski, H. L. Liber, T. R. Skopek, S. A. Slapikoff, R. J. Tizard and B. W. Penman, in "Chemical Mutagens," Vol. 6, F. J. DeSerres and A. Hollaender, eds., Plenum Press, New York, 1980, p. 331.

20. E. E. Furth, W. G. Thilly, B. W. Penman, H. L. Liber and W. M. Rand, Quantitative assay for mutation in diploid human lymphoblasts using microtiter plates, Analyt. Biochem., 110, 1 (1981).

21. H. L. Liber and W. G. Thilly, Mutation assay at the thymidine kinase locus in diploid human lymphoblasts, Mutat. Res., 94, 467 (1982).

22. S. A. Slapikoff, B. M. Andon and W. G. Thilly, Comparison of methylnitrosourea, methylnitrosoguanidine and ICR-191 among human lymphoblast lines, Mutat. Res., 70, 365 (1980).

23. T. R. Skopek, H. L Liber, D. A. Kaden and W. G. Thilly, Relative sensitivities of forward and reverse mutation assays in Salmonella typhimurium, Proc. Natl. Acad. Sci. (USA), 75, 4465 (1978).

24. C. Wayhas, K. Weigl and H. Sies, The disposition of formaldehyde and formate arising from drug N-demethylations dependant on cytochrome P-450 in hepatocytes and in perfused rat liver, Europ. J. Biochem., 89, 143 (1978).

25. W. D. Tephly, W. D. Watkins and J. I. Goodman, The biochemical toxicology of methanol, Essays Toxicol., 5, 149 (1974).

26. N. Y. Feldman, Reactions of nucleic acids and nucleoproteins with formaldehyde, Prog. Nucleic Acid Res. Mol. Biol, 13, 1 (1973).

27. B. W. Penman, C. L. Crespi, E. A. Komives, H. L. Liber and W. G. Thilly, Mutation of human lymphoblasts exposed to low concentrations of chemical mutagens for long periods of time, Mutat. Res. (In Press).

DISCUSSION

DR. STERNBERG: (Stephen Sternberg, Memorial Cancer Center, Sloan Kettering.) Did you test formaldehyde in pure form; that is, in the absence of methanol?

DR. GOLDMACHER: No. It is very difficult to do because formaldehyde in solution is very unstable. We did use methanol as a negative control in both bacterial and human cell assays. So, it is pretty clear that it didn't cause any mutations or any toxicity at all.

DR. STERNBERG: But I wonder if the combination of the two may be the important thing.

DR. GOLDMACHER: Yes, it might be.

DR. BRUSICK: (David Brusick, Litton Bionetics.) To answer that question, we did some work with mouse lymphoma cells, not human cells, starting with paraformaldehyde and putting it in aqueous solution. It was also mutagenic at the TK locus, and the kinetics of mutation induction for TK mutation in mouse lymphoma cells were almost identical to what you showed in human cells. The lowest effect level was roughly four or five parts per million, and the same kind of curve was obtained. So, it is an effect of the formaldehyde and not of the combination of methanol and formaldehyde.

DR. GOLDMACHER: That is very interesting. Have you published those results?

DR. BRUSICK: I believe that they are in some publication. I don't know if it is in the book (Ed. note: Proceedings of the 1980 Formaldehyde Conference) that is coming out. It has been presented at at least two meetings.

DR. CLARY: They are in the CIIT book of the 1980 Conference.

DR. BERNSTEIN: (Martin Bernstein, Ciba-Geigy.) Is there any relationship between the concentrations you used _in vitro_ and what someone could be exposed to, either by inhalation or dermally? Second, could you have exposed these cells in a closed system to ppm levels of formaldehyde gas?

DR. GOLDMACHER: It is very difficult to answer these questions. I tried to find the formaldehyde partition coefficient between air and water in the literature, but I could not find it. I don't know, if you have, for example, 4 ppm in air, what the concentration will be in the water. So, I cannot answer your question.

DR. BRUSICK: (David Brusick, Litton Bionetics.) I might be able to answer that. Again, we tried to make some calculations on mouse lymphoma data along the same lines, and the lowest effective concentration probably would be at levels higher than normally encountered by human exposure. I am not sure, but maybe somewhere between 15 and 20 ppm by inhalation would be close to the level at which we saw mutation induction in an _in vitro_ system.

DR. GOLDMACHER: How did you do your assay?

DR. BRUSICK: The assay was done in essentially identical fashion to the human cell assay -- *in vitro* with aqueous exposure.

DR. GOLDMACHER: Yes, but what did you just say about exposure in air?

DR. BRUSICK: The calculations were based upon the normal inhalation volume of a mouse and the solubility of formaldehyde and some very crude calculations. We tried to get the dose per cell in the target organ tissue based upon inhalation exposure and then compared that on a milligram per kilogram body weight basis to an *in vitro* system in which you just put the material in an aqueous solution. So, the target cell dose that was mutagenic *in vitro* would be a little higher than the typical human inhalation exposure. But, it was in the range of the doses that were used in the *in vivo* carcinogenicity inhalation studies that were conducted by CIIT.

DR. GOLDMACHER: It is very difficult to decide what the body fluid concentration will be after inhaling formaldehyde. First of all, formaldehyde solubility in water is very high, and secondly, we don't know the partition coefficient. So, it might be 100 times higher or 100 times lower, *in vivo* vs. *in vitro*.

DR. BRUSICK: Yes, they were intended just as very crude estimates and not as accurate determinations.

DR. BEALL: (James Beall, Department of Energy.) Having read some other papers put out by your group and Dr. Thilly, I have read from time to time that he has found that in the lower ranges of his testing system that multiple doses at these lower concentrations will often give the same effect as one dose at a higher concentration. Because of this, it raises the issue of whether, in fact, multiple doses at lower than say, 4 parts per million, might give you the same mutagenic effect in your human cell line as the dose that you gave, at least 4 parts per million. Would you philosophize on that for a second?

DR. GOLDMACHER: This was shown for some alkylating mutagens *in vitro*. To extrapolate to an *in vivo* situation is too difficult; I will not do it.

DR. BEALL: I thought you might elaborate on the *in vitro* assays you have done which show that.

DR. GOLDMACHER: As for *in vitro* assays, it depends. If you have, for example, any serial system which will allow cell recovery after the first dose, you will have no additive effect from the second dose and so on. If you have no such system, for example, DNA repair system, the effect will accumulate.

In our cell line, for example, when we treated cells with repeated doses of alkylating carcinogens, the number of mutations rose and rose and rose, and this is because our particular line evidently has no DNA repair system against mutagenic carcinogens. As for lesions caused by formaldehyde? Nobody knows. I do not know what kind of lesions in DNA are caused by formaldehyde, and whether there are any repair systems or not. It is impossible to say about formaldehyde.

DR. HECK: (Henry Heck, CIIT.) It is very difficult to compare your type of experiment with an *in vivo* experiment because in your experiment you do, or you can, measure free formaldehyde. In an *in vivo* experiment, you can measure cellular formaldehyde, and the result of our analyses, which I will mention this afternoon, were that there were no changes in the cellular formaldehyde concentrations in the nasal mucosa. So, it is quite hard to make a direct comparison with an *in vivo* result.

DR. GOLDMACHER: That is right, and I have never compared concentrations of formaldehyde in our experiments with *in vivo* concentrations.

DR. BRUSICK: (David Brusick, Litton Bionetics.) After the previous discussions of the initiation-promotion, I would like to make a comment. It deals with my impression that for the mouse, formaldehyde is not an initiator, and it may or may not be a promoter. If it is a promoter in the mouse, it is very, very weak. I think we have to keep in mind that we are dealing with a single unique species, and results in that species cannot be interpreted as indicators that formaldehyde is not an initiator or promoter in another species, such as the rat. We know that many chemicals which are carcinogens are often carcinogens in a single species, meaning that they are either initiators or promoters in that species and not in another.

For example, in the bioassay program at NCI, up to as much as 30 percent of the chemicals are carcinogenic only in a single species.

DR. CLARY: I think that point is well taken.

CHAPTER 8

FORMALDEHYDE AND THE NASAL MUCOCILIARY APPARATUS

K. T. Morgan, D. L. Patterson and E. A. Gross

Chemical Industry Institute of Toxicology
Research Triangle Park, North Carolina

The nasal mucociliary apparatus serves to remove inhaled particles deposited in the nasal passages, and may also act as a protective barrier against inhaled gaseous irritants. The investigations described are aimed at assessing the protective role played by the mucociliary apparatus during exposure to formaldehyde gas. A system has been designed to study *in vitro* mucociliary responses during *in vitro* exposure, as well as following *in vivo* exposure. Initial investigations using the frog palate, which is a widely used model for studies of mucociliary function, demonstrated clear concentration-related responses to formaldehyde. There was initial stimulation of mucus flow rate due to increased ciliary activity, and then subsequent inhibition of flow, which was attributed to a direct effect of formaldehyde on the superficial mucous layer. Studies in rats and mice indicate that most of the nasal cavity is covered by a continuous layer of mucus, and there are very similar and specific patterns in the two species. Mucociliary function of rodent nasal passages appears to be more resistant to *in vitro* formaldehyde exposure than that of the frog palate. The potential of the mucociliary apparatus to protect the upper respiratory tract against formaldehyde toxicity is discussed.

INTRODUCTION

The respiratory tract presents a number of potentially protective barriers or responses to inhaled irritant gases such as formaldehyde, including reduction of minute volume[1], the mucociliary apparatus[2], the cell membrane and intercellular junctions, and cellular metabolism[3]. The nasal squamous cell carcinomas induced in rats by formaldehyde occurred in areas normally lined by ciliated respiratory epithelium[4] which is covered by a superficial layer of mucus. A great deal of information has been accumulated on mucociliary function in the tracheo-bronchial tree of mammals[2] and the response of mucociliary systems to formaldehyde (Table 1), but much less is known about the role of this system in the nasal passages[5]. The purpose of these investigations was to study normal mucociliary function of the nasal passages of rats, and subsequently to assess the effects of formaldehyde gas upon mucociliary activity. The relevance of the nasal mucociliary apparatus to potential thresholds for formaldehyde toxicity are discussed.

METHODS

Animals

Male F-344 rats were obtained from Charles River Breeding Laboratories Inc. (Kingston, NY). Upon arrival, all animals were placed in quarantine, housed in groups of 3 in stainless steel cages and received food (NIH-07, Ziegler Bros., Gardners, PA) and tap water *ad libitum*. Rats were all in a good state of health

TABLE 1

Formaldehyde Effects on the Mucociliary Apparatus

Species (Ref. Number)	Concentration	Effect
Rabbit Trachea [7]	30-60 ppm	Ciliastasis
	60-100 ppm	Ciliastasis
Rat Trachea [8]	22 ppm	Mucostasis and Ciliastasis
	10 ppm	"
	3 ppm	"
	0.5 ppm	"
Frog Esophagus & Rat & Rabbit Trachea[9]	18,000-54,000 ppm	Immediate mucostasis and ciliastasis
Frog Esophagus[10]	1-3 mg*	Deceleration of beat frequency
Cat Trachea[11]	Total of 10 μg^{++} inhaled	Up to 50% reduction in mucus flow rate
	Total of 20 μg inhaled	Up to 58% reduction in mucus flow rate
Rabbit Trachea**[12]	6-48 μg^{+}	Ciliary inhibition
Human Nose[13]	0.38-1.63 ppm	Decrease in mucus flow rate

* Applied as a rinse
**Immersed tracheal preparation
$^{+}$ In puffs of 12 sec duration
$^{++}$Variable number of puffs

and had negative virus titers in the standard murine virus antibody determination (Microbiological Associates, Bethesda, MD).

In Vitro Analysis System

The equipment used for these studies has been reported previously[6] and will be described only briefly. The system consists of a 140 ml gas-tight chamber equipped with gas humidification, mixing, delivery and analytical sampling systems and a water jacket for temperature control. Ultra-zero[R] air (Matheson) was delivered to the chamber from a cylinder at a pressure of 6 psi. The air flow was controlled with flow meters and humidified by passing part or all of the air-stream through distilled water at room temperature using a fritted bubbler. Humidity in the chamber was assessed by measuring the dew-point of the exhaust air stream from the chamber with a dew-point probe and hygrometer and maintained at 70%-80%. The chamber flow rate was maintained between 100 and 1030 ml per minute, and was measured with a glass bubble meter in the exhaust stream. The chamber was maintained at a temperature of 37°-38°C.

The surface to be studied was illuminated with a fiber-optics light source using a steep incident light angle. The microscope was fitted with long working distance objectives, and a trinocular head. Recordings of the image were made with a Sony Rotary Shutter Television Camera, with a shutter speed of 60 frames per second. The signal from the camera passed through a time/date generator, with built in stop watch using 100th second units, and a Video Motion Analyzer, and was recorded on video magnetic tapes with a video cassette recorder having image scanning capability. The video motion analyzer permitted controlled speed or frame-by-frame examination.

In Vivo Exposures

Rats were exposed to formaldehyde gas in an 8 cubic meter glass and stainless steel chamber, with the test atmosphere generated as previously described[1]. Rats were killed by rapid decapitation; treated animals were killed as soon as possible (always less than 1 hour) following cessation of formaldehyde exposure. The head was skinned and lower jaw and excess tissue removed. The palate was incised along the midline to a depth of about 2 mm and the skull split in half along the sagittal suture from the coronal suture to the tip of the nasal bones. The nasal septum was removed carefully from the right half of the head to expose the turbinates. The nasoturbinate was rapidly removed from the left side and placed with the right half in the exposure chamber. Short recordings (10-15 sec) were made in a consistent manner for each area to be studied and flow rates were determined from these recordings. Speed of recording was very important for this procedure as normal mucus flow was fairly short lived. (20-40 minutes).

Data Capture

Recordings were made at the standard locations shown in Fig. 1. Mucus flow rate (mm/min) was measured by timing a calibrated distance travelled by particles present in the superficial mucous layer. Ciliastasis was defined as the absence

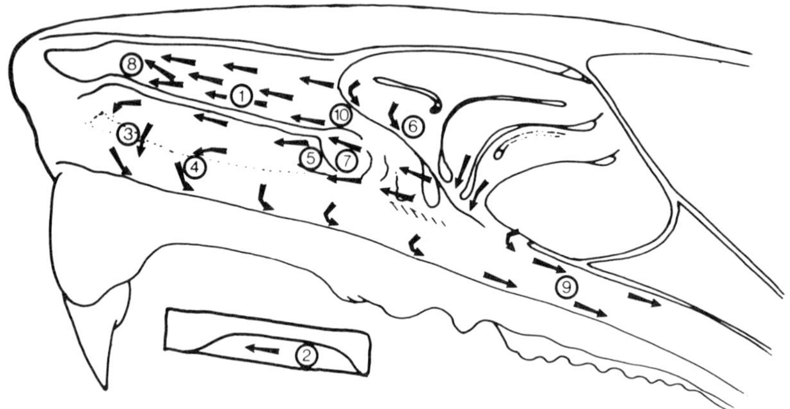

Fig. 1. Mucus flow pattern in rat nasal passages. Numbers 1-10 indicate areas used for video recordings. Area 2 is on the lateral aspect of the left nasoturbinate which was dissected free and examined with the right half of the skull.

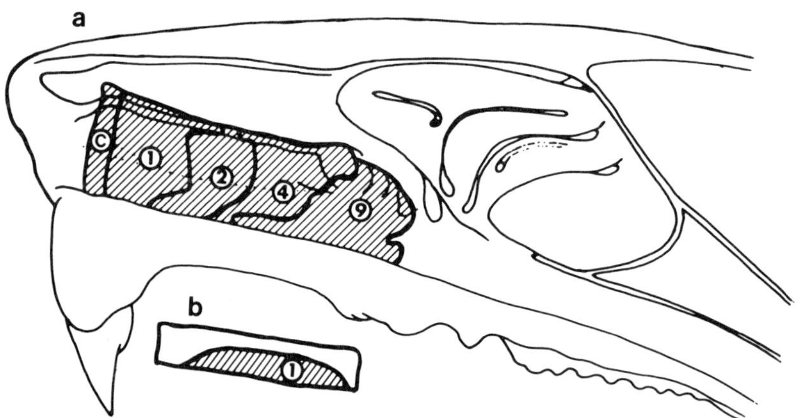

Fig. 2. a. Maps indicating areas of defective mucociliary function based upon subjective interpretation. Progression within nasal cavity of areas of mucostasis and partial to complete ciliastasis with increasing numbers of days of formaldehyde exposure (15 ppm; 1, 2, 4 or 9 days, 6 hrs. per day). "C" indicates the anterior area of respiratory epithelium in which mucus flow and ciliary activity are not apparent in control animals.

b. Left nasoturbinate showing lateral aspect and presence of mucostasis and ciliastasis from Day 1.

of ciliary activity in areas in which cilia were beating in control animals. Mucostasis and/or ciliastasis were recorded on maps of the nasal passages (Fig. 2).

RESULTS

Control Rats

Mucus was seen to be present as a flowing continuous coat over the respiratory epithelium except on the most anterio-ventral extremity of the nasoturbinate and the anteriomedial extremity of the maxilloturbinate (Fig. 2). Normal flow patterns observed in all control animals are presented in Fig. 1, which also shows the areas used for video analysis. The mean flow rates (± S.E.) for the ten areas are presented in Table 2, with time required to make these analyses and details of ciliary activity and mucus flow. Most of the flow was anterior except in the ventral nasal passages and the anterior aspect of the ethmoid turbinates on which flow was towards the orifice of the nasopharynx (Fig. 1). In most areas of the nose mucus flow persisted for at least 20 minutes, while in the nasopharynx it consistently lasted for more than 30 minutes. The mean time to analysis and mucus flow rates for the ten areas recorded are in Table 2. Flow was fastest in the nasopharnyx, slightly slower on the posterior lateral wall and mid-region of the nasoturbinate, and consistently slow on the anterioventral tip of the naso-turbinate. The anteriomedial aspect of the maxilloturbinate (Fig. 2), just posterior to the squamous epithelium of the vestibule, exhibited no flow but appeared to be covered by mucus,

TABLE 2

Mucociliary Function in Control Animals

	Areas of Nasal Passages*				
	1	2	3	4	5
Time to Data Capture ± SE (n) in minutes	6.95 ± 0.94 (8)	7.28 ± 0.23 (8)	9.19 ± 0.60 (9)	9.67 ± 0.29 (6)	10.16 ± 0.40 (8)
Mucus Flow⁺ Rate ± SE (n) in mm/minute	5.88 ± 0.46 (8)	0.28 ± 0.57 (8)	1.02 ± 0.53 (9)	1.48 ± 0.59 (6)	3.63 ± 0.80 (8)
Number Exhibiting** Mucus Flow/Number Examined	8/8	1/8	9/9	7/7	10/10
Number Exhibiting Ciliary Activity/Number Examined	8/8	8/8	9/9	7/7	10/10
	6	7	8	9	10
Time to Data Capture ± SE (n) in minutes	12.1 ± 0.36 (9)	13.41 ± 0.77 (7)	13.47 ± 0.49 (9)	15.55 ± 0.62 (9)	19.85 ± 0.92 (5)
Mucus Flow⁺ Rate ± SE (n) in mm/minute	0.36 ± 0.24 (8)	4.92 ± 1.81 (7)	1.13 ± 0.34 (9)	9.02 ± 0.85 (6)	2.64 ± 1.16 (5)
Number Exhibiting** Mucus Flow/Number Examined	4/9	6/7	9/9	9/9	5/5
Number Exhibiting Ciliary Activity/Number Examined	9/9	7/7	9/9	9/9	5/5

* See Fig. 1.
**In some animals flow was observed but it was not possible to determine flow rate.
⁺ Animals with no flow or very slow flow were used as zero values for these calculations.

and had variable degrees of drying or crusting of the surface. The lateral aspect of the nasoturbinate had mucus flow in only one case but all controls exhibited vigorous beating by the sparse population of cilia in this location. Areas of slow flow were generally poorly populated by ciliated cells. No evidence of mucus flow could be detected in areas of olfactory epithelium.

Formaldehyde Exposed Rats

Formaldehyde exposure induced mucostasis and ciliastasis in specific areas of the nasal passages with very consistent responses within each exposure group, and the extent of these changes increased with increasing number of exposure days. The distribution of areas of ciliastasis and associated mucostasis is shown in Fig. 2. This effect of formaldehyde was most severe in the anterior nasal passages in areas which were sparsely populated by ciliated cells. The results of video analyses are presented in Table 3. It is evident from Table 3 that mucostasis was present in almost all animals in Area 2, and in more than half of the animals in Area 6. The nasoturbinate had a very sparse ciliated cell population and was removed as a small fragment, both of which factors may have contributed to limited *in vitro* flow in this region. In Area 6 (anterior ethmoid margin) there was vigorous ciliary activity but mucus flow had stopped in most cases. This region has been found to stop flowing sooner than other areas of the nasal passages and would best be examined sooner in future studies. Mucostasis preceded the induction of ciliastasis in most areas affected. In areas

TABLE 3

Videoanalysis of Mucostasis and Ciliastasis

Number with Mucostasis/Number examined

Treatment Group**	Area of Nasal Passages* 1	2	3	4	5
Control	0/8	7/8	0/9	0/7	0/10
1 d HCHO	0/4	1/1	4/4	4/4	4/4
2 d HCHO	0/4	2/2	3/4	4/4	3/4
4 d HCHO	1/4	3/3	4/4	4/4	3/4
9 d HCHO	0/4	3/3	4/4	4/4	4/4

	6	7	8	9	10
Control	5/9	1/7	0/9	0/9	0/5
1 d HCHO	2/4	1/2	3/3	0/3	0/2
2 d HCHO	2/3	ND	3/4	0/4	1/3
4 d HCHO	3/4	2/2	4/4	0/4	3/4
9 d HCHO	2/4	4/4	4/4	0/4	1/3

Number with Ciliastasis/Number examined

Treatment Group**	Area of Nasal Passages* 1	2	3	4	5
Control	0/8	0/8	0/9	0/7	0/10
1 d HCHO	0/4	1/1	1/4+	1/4+	0/4
2 d HCHO	0/4	2/2	1/4+	0/4+	0/4+
4 d HCHO	0/4	3/3	4/4	4/4	3/4
9 d HCHO	0/4++	3/3	4/4	4/4	4/4

	6	7	8	9	10
Control	0/9	0/7	0/9	0/9	0/5
1 d HCHO	0/4	0/2	0/3	0/3	0/2
2 d HCHO	0/4	ND	2/4+	0/4	0/3
4 d HCHO	0/4	1/2	3/4	0/4	0/4
9 d HCHO	0/4	4/4	4/4	0/4	0/3

* See Fig. 1.
**Formaldehyde exposures were at 15 ppm for 6 hours per day.
\+ Much reduced ciliary activity.
++Ciliastasis on margin of turbinate.
ND = not done

where flow persisted there was some slowing, but small group sizes limited the value of statistical analysis of flow rate data.

DISCUSSION

The finding that formaldehyde induced squamous cell carcinomas in the nasal passages of rats contributed to the restrictions imposed on the sale of urea-formaldehyde foam insulants[14] and raised considerable controversy with respect to future regulation of formaldehyde[15]. One important issue in the regulation of potential human carcinogens is the concept of a threshold or no-effect level. Although considerable effort has been expended in this regard, with current methods threshold or no-effect levels cannot be reliably established for the human population[16].

The present studies indicate that in the rat nasal passages the mucus is continuous over most of the respiratory epithelium. Thus, before chemicals or particles can directly affect the underlying epithelium they must penetrate this superficial layer. Mucus consists principally of water ($\geq$ 95%), mucous glycoproteins (0.5-1%), free proteins and salts, with other materials in much smaller amounts[17]. Formaldehyde binds readily to proteins[18] and also reacts with polysaccharides at room temperature[19]. These reactions, if combined with constant mucus removal and replacement, might be expected to impede penetration of formaldehyde to the periciliary fluid layer. Furthermore, if the rate of binding and removal of formaldehyde by the mucous layer were equal to the rate of delivery of the gas to the exposed surface, then one could reasonably assume that the underlying epithelium would not be exposed. If it is assumed that formaldehyde must reach the epithelial cells before it induces any neoplastic response, then assessment of the effectiveness of the mucus of other potentially protective 'barriers' may permit estimation of a threshold or no-effect level.

There are many ways of studying mucociliary function in the airways[20] but due to the small size and narrow, tortuous nature of rat nasal passages it is not yet possible to obtain detailed information concerning mucus flow patterns and flow rates *in vivo*. Studies of tracheal mucus velocity in a range of animal species demonstrated no consistent differences between *in vitro* and *in vivo* measurements[21]. The nasal mucus flow rates found in rats were in the same range as those reported for the human nasal passages[5] which suggests that these *in vitro* measurements may reflect the *in vivo* situation.

Only one published report of rodent nasal mucus flow patterns was found[22], and these authors described similar flow patterns to those presented here with some minor differences. The anteriorly directed flow present in most of the nose would appear to function as a 'countercurrent' flow system with respect to imcoming (contaminated) air. An area of anterior flow, the 'clearance current' has also been reported in human nasal passages, and this is an area where most of the inhaled particles are deposited[5] and may represent a point of similarity between human and rodent nasal mucociliary function.

Studies of *in vitro* responses by the frog palate mucociliary apparatus to formaldehyde gas[6] demonstrated a clear concentration relationship for inhibition of mucus flow across the concentration range used in the chronic formaldehyde inhalation toxicity study in rats and mice (15, 6 or 2 ppm)[4]. At the higher concentration, mucostasis and ciliastasis were rapidly induced in the frog palate. At 1.4 ppm and below, these effects were not induced and mucociliary function was apparently unimpaired. If a similar concentration relationship were found for impairment of nasal mucociliary function in rats and mice then the protective action of the flowing mucous layer would be inactivated at 15 and possibly 6 ppm. At lower concentrations the mucociliary apparatus would continue to remove formaldehyde and reduce exposure of the underlying cells.

It has been established that formaldehyde gas can impair mucociliary function, both *in vitro* and *in vivo* in man and animals (Table 1). These workers have used widely differing exposure systems, test tissues and exposure regimens. It can be seen in Table 1 that there were varying results concerning both the time and concentration of formaldehyde required to induce ciliastasis in the different test systems. Dalham et al[22] were the only workers to apply formaldehyde concentration over the range used for the chronic toxicity in rats and mice[4] and they found that 10 ppm formaldehyde induced ciliastasis in the opened trachea of anesthetized rats in 10 seconds. It would appear from the present study that the nasal passages of the rat are much more resistant to the ciliastatic activity of formaldehyde than this during *in vivo* exposure.

The current studies in rats demonstrated clearly localized impairment of mucociliary function following *in vivo* exposure to 15 ppm. Mucostasis preceded ciliastasis, as was found for acute, *in vitro* formaldehyde exposure using the frog palate preparation[6]. The most severely affected areas included the naso- and maxillo-turbinates and adjacent lateral wall, which are regions of the nose in which histological changes are first detected following acute inhalation exposure[23]. It is thus possible that the distribution of formaldehyde-induced lesions is related to varying effectiveness of different regions of the nasal mucociliary apparatus. Work is in progress to study the effects of lower concentrations of formaldehyde on nasal mucociliary function. It has been shown recently that the mucous lining of the intestine provides a diffusion barrier to nutrients[24]. Continued studies on the role of respiratory tract mucus as a barrier to inhaled irritant gases may be useful in assessing its role in the protection of the underlying cell population from potential carcinogens such as formaldehyde.

ACKNOWLEDGMENTS

We thank Dr. Craig Barrow, Dr. J. A. Swenberg, Mr. Mark Phelps, Mr. William Steinhagen, Mr. Richard James, and Dr. X. Z. Jiang for considerable advice, help, and encouragement with this work, and Ms. Holly Randall for histological work.

REFERENCES

1. J. C. F. Chang, E. A. Gross, J. A. Swenberg and C. S. Barrow, Comparison between B6C3F1 mice and F-344 rats on nasal cavity deposition, histopathology and cell proliferation following single and repeated formaldehyde exposures. Toxicol. Appl. Pharmacol. In Press.

2. A. Wanner, Clinical aspects of mucociliary transport. Am. Rev. Respir. Dis., 116, 73 (1977).

3. M. Casanova-Schmitz and H.d'A. Heck, Metabolism of formaldehyde in the rat nasal mucosa *in vivo*. The Toxicologist, 3, 61 (1983).

4. J. A. Swenberg, W. D. Kerns, R. E. Mitchell, E. J. Gralla and K. L. Pavkov, Induction of squamous cell carcinomas of the rat nasal cavity by inhalation exposure to formaldehyde vapor. Cancer Res., 30, 3398 (1980).

5. D. F. Proctor, The mucociliary system. In: "The Nose, Upper Airway Physiology and the Atmospheric Environment", p 245, D. F. Proctor and I. Andersen, Eds., Elsevier, Amsterdam, 1982, p. 245.

6. K. T. Morgan, D. L. Patterson, and E. A. Gross, Frog palate mucociliary apparatus: structure, function and response to formaldehyde gas. Lab. Invest. Submitted.

7. L. Cralley, The effect of irritant gases upon the rate of ciliary activity. J. Ind. Hyg. Toxicol., 24, 193 (1942).

8. T. Dalhamn, Mucous flow and ciliary activity in the trachea of healthy rats and rats exposed to respiratory irritant gases. Acta Physiol. Scand., 36, suppl. 123 (1956).

9. P. Kotin, Unpublished results (1957) cited by: H. E. Stokinger and D. L. Coffin., Biologic effects of air pollutants, In "Air Pollution," 2nd Ed., A. C. Stern, Ed., Academic Press, NY, 1968, p. 483.

10. H. L. Falk, P. Kotin, and W. Rowlette, The response of mucus-secreting epithelium and mucus to irritants. Ann. N. Y. Acad. Sci., 106, 583 (1963).

11. S. Carson, R. Goldhamer, and R. Carpenter, Responses of ciliated epithelium to irritants, Am. Rev. Respir. Dis., 93, 86 (1966).

12. C. J. Kensler and S. P. Battista, Chemical and physical factors affecting mammalian ciliary activity. Am. Rev. Respir. Dis., 83, 93 (1966).

13. I. Anderson and L. Molhave, Controlled human studies with formaldehyde in "Proceedings of the Third CIIT Conference in Toxicology: Formaldehyde Toxicity", J. E. Gibson, Ed., Hemisphere, Washington, D.C. (In Press).

14. Consumer Product Safety Commission, Fed. Regist. 47, 14366.

15. F. Perera and C. Petito, Formaldehyde: A question of cancer or policy, Science, 216, 1285 (1982).

16. Office of Technology Assessment, Assessment of technologies for determining cancer risks from the environment. OTA, Washington, D.C., June, 1981.

17. J. M. Creeth, constituents of mucus and their separation. Br. Med. Bull., 34, 17 (1978).

18. D. French and J. T. Edsall, The reactions of formaldehyde with amino acids and proteins. Adv. Protein Chem., 2, 277 (1945).

19. Y. Kihara, M. Kasuya, and K. Tanaka, Reaction of aldehydes on starch. I Dempun Kogyo Gakkaishi, 10, 1 (1962).

20. R. J. Phipps, The airway mucociliary system., in "Int. Rev. Physiol., Resp. Physiol" III, 23, J. G. Widdicombe, Ed., University Press, Baltimore, 1981.

21. T. Asmundsson and K. H. Kilburn, Mechanisms of respiratory tract clearance. In "Sputum, Fundamentals and Clinical Pathology," M. J. Dulfano, Ed., C. C. Thomas, Springfield, IL (1973).

22. A. M. Lucus and L. C. Douglas, Principles underlying ciliary activity in the respiratory tract. II. A comparison of nasal clearance in man monkey and other mammals. Arch. Otolaryngol., 20, 518 (1934).

23. J. A. Swenberg et al., Mechanisms of formaldehyde, in "Proceedings of the Third CIIT Conference in Toxicology: Formaldehyde Toxicity", J. E. Gibson, Ed., Hemisphere, Washington, D.C. (In Press).

24. D. B. Millar, L. R. Jacobs, and G. M. Gray, Intestinal diffusion barrier: Unstirred water layer or membrane surface mucous coat? Science, 214, 1241 (1981).

DISCUSSION

DR. STERNBERG: (Stephen Sternberg, Memorial Cancer Center, Sloan Kettering.) I want to congratulate Dr. Morgan for a very elegant presentation. I would like to emphasize some points he has made, mainly the very large nose of rodents and the very large surface area that the turbinates encompass. If you can imagine a proportionately large nose on a human, you would realize how little effect we would have breathing in formaldehyde and other items nasally.

The other point is that the rodent is an obligate nasal breather. It cannot breathe through its mouth, and I think it is worth remembering that people exposed to formalin or formaldehyde may start breathing through their mouth when they smell it. Thus, it may be the oral cavity and the oral pharynx that should be examined in man for tumors, rather than the nasopharynx.

DR. MORGAN: With respect to nasomucociliary function, Dr Proctor has written an excellent book on the human nose. In man, relative to the size of the head, there is a much-reduced olfactory area. I did not talk about olfactory function here. What I am basically talking about here is the reason for the effect on the rat. I tried to focus the presentation on animals, not humans, except for the pictures you saw.

CHAPTER 9

REACTIONS OF FORMALDEHYDE IN THE RAT NASAL MUCOSA

H. d'A. Heck and M. Casanova-Schmitz

Chemical Industry Institute of Toxicology
Research Triangle Park, North Carolina

Possible reactions of formaldehyde (HCHO) in the nasal mucosa include the formation of reversible adducts and irreversible cross-links (which may or may not be toxic) and metabolism. In this case, metabolism is primarily a mechanism of detoxification. Exposure of Fischer 344 rats to HCHO at 15 ppm did not cause a significant change in the total concentration of free and reversibly bound HCHO in the nasal mucosa. However, evidence was obtained that adducts reducible with [^{3}H]$NaBH_4$ were formed with nasal mucosal DNA. Moreover, formaldehyde decreased the amount of DNA that was extractable from proteins under denaturing conditions, whether the reaction of HCHO with the nasal mucosa was performed *in vitro* under mild conditions or resulted from an *in vivo* exposure at 15 ppm (6 hr. daily for 2 days). Despite the changes in the extractability of the DNA, no apparent change was noted in the density of the DNA as measured by isopycnic centrifugation in CsCl. These results indicated that HCHO forms DNA adducts and DNA-protein cross-links in the rat nasal mucosa at 15 ppm.

INTRODUCTION

Formaldehyde (CH_2O) reacts readily with nucleophiles, including primary and secondary amines, thiols, hydroxyls, and amides, to form reversible adducts. These compounds contain one or two active hydrogens, and the reaction product is a methylol derivative. Primary amines can bind two molecules of CH_2O to form essentially stable cross-links.

It is important to emphasize the reversibility of the initial reactions. Many dissociation constants for the reaction of CH_2O with amino acids, peptides, nucleosides, and nucleotides have been measured.[1-3] The nucleophilic group is usually an amine in these reactions. Although the full range of constants covers about three orders of magnitude, most of the values were between 2 and 200 mM. Given the low concentrations of CH_2O that seem to exist endogenously in biological tissues (i.e., 0.05 to 0.5 μmole/g wet weight of tissue),[4] any particular cellular nucleophile having a dissociation constant in this range would necessarily have a low probability of occupation, and the corresponding reversible adduct would have a short lifetime. Moreover, reversibly bound CH_2O would be expected to equilibrate rapidly within the intracellular matrix and extracellular fluids. To the extent that inhaled CH_2O is reversibly bound, it should not differ from endogenous CH_2O in these respects. Direct evidence that inhaled CH_2O equilibrates with the endogenous CH_2O pool in the nasal mucosa of Fischer-344 (F-344) rats at both 2 and 15 ppm has been obtained by use of a double-labeling method.[5]

METHOD, RESULTS AND DISCUSSION

The concentration of free and reversibly bound CH_2O in the nasal mucosa of F-344 rats did not increase significantly following repeated exposures to toxic CH_2O concentrations (These results are presented in Table 1). The free and reversibly bound CH_2O was analyzed by gas chromatography/mass spectrometry.[4]

Reversible binding of CH_2O was also investigated by another method in which the nasal mucosa of F-344 rats exposed to 15 ppm of CH_2O was reacted with $[^3H]NaBH_4$.[6] The tissue was dissected as soon as possible after exposure, and the reactions were carried out under conditions that were shown *in vitro* to cause extensive labeling of proteins, RNA, and DNA with tritium by the reduction of reversible CH_2O (Schiff base) adducts to stable methylamines. No statistically significant increase in tritium labeling of proteins and RNA was observed in the exposed animals in comparison with controls (Table 2). However, a significant increase in the tritium content of the DNA was apparently obtained.

TABLE 1

Concentrations of Formaldehyde in the Nasal Mucosa of Control (i.e., Unexposed) F-344 Rats and of Rats Exposed to Formaldehyde*

Average Formaldehyde concentration** (μmoles/g wet wt.)			Result of ANOVA test for among-sample averages
0 ppm	6 ppm	15 ppm	
0.34 ± 0.032 (16)	0.385 ± 0.041 (8)	0.353 ± 0.036 (8)	NS ($p > 0.5$)

* Exposure duration: 6 hr./day, 10 days.
**Mean ± S. E.; No. of samples is shown in parentheses.

TABLE 2

Labeling of Proteins, RNA, and DNA with Tritium after Incubation of Nasal Mucosal Homogenates from Unexposed Male F-344 Rats and from Rats Exposed to CH_2O (15 ppm, 6 hr. daily for two days) with $[^3H]NaBH_4$*

Macromolecular Fraction+	Specific activity (DPM/mg)**		Significance Level++
	Control	Exposed	
Protein	2686 ± 395	2910 ± 215	NS (0.25 > p > 0.10)
RNA	797 ± 249	933 ± 287	NS (0.25 > p > 0.10)
DNA	135 ± 35	220 ± 48	Sig (0.01 > p)

* Respiratory mucosa was homogenized and incubated for 30 min. in the presence of $[^3H]aBH_4$ (7.5 mM; 5.0 μCi/mmole). Incubation conditions: 0°C, 0.1 M phosphate containing 5 mM EDTA (pH 8.0). The tissue was solubilized using 8 M urea, 0.24 M sodium phosphate, 10 mM EDTA, and 1% sodium dodecylsulfate (pH 6.8) and was extracted with chloroform/iso-amyl alcohol/phenol (24/1/25).

**Values shown are means ± S.D. for five groups of control and exposed rats containing three animals per group.

+ Protein was recovered from the organic phase of the extraction by precipitation with acetone and was washed with acetone, ether, and 95% ethanol. The proteins were then dialyzed against 2 x 2 L of 0.01 M NaOH. Nucleic acids were purified by hydroxyapatite chromatography and were concentrated and washed with water in a stirred ultra-filtration cell. The specific activities were determined by scintillation counting and are corrected for background using appropriate blanks in all cases.

++One-tailed *t*-test; NS = Not Significant; Sig = significant.

As the total counts in the DNA were very low (only 50% greater than background), the increase in DNA radioactivity should be regarded with caution. The data are consistent with the hypothesis that Schiff base adducts were formed at very low concentrations in the DNA of the rat nasal mucosa following exposures to 15 ppm of CH_2O, but they should not be regarded as conclusive. It is evident that the data do not exclude the possibility that similar adducts were formed with proteins and RNA, but were undetectable owing to the higher backgrounds in those samples.

An as yet unexplained aspect of the preceding experiments is the high specific activities measured in the proteins, and to lesser extents, in the RNA and DNA of control animals. It has not escaped our attention that these counts might be caused in part by $[^3H]NaBH_4$ reduction of endogenous CH_2O or other aldehydes convalently bound to amine moieties in the macromolecules. This possibility deserves further consideration, as the results could pertain directly to the possible existence of a biological threshold for CH_2O exposure.

Although adduct formation with amines is a reversible reaction, it can lead to cross-links which _are_ stable, do not equilibrate, and may require special mechanisms for their removal. Cross-linking that involves critical, probably macromolecular, cellular components can induce a variety of toxic responses if repair mechanisms are inadequate to deal with the products formed.

Several studies have shown that CH_2O forms cross-links between DNA and proteins _in vitro_. Cross-linking in isolated nucleohistones has been extensively investigated.[7-10] This reaction occurs rapidly at relatively low concentrations of

CH_2O. Cross-linking within or between strands of isolated DNA has also been observed, but this reaction seems to occur much more slowly.[11] Cross-linking of DNA to proteins by CH_2O has also been detected *in vivo* in microorganisms such as *E. coli*[12] and yeast,[13] and *in vitro* in mammalian cells.[14,15]

Whether cross-linking of DNA to proteins is directly responsible for the mutagenicity of CH_2O in cell cultures has not been established.[16,17] However, it has been shown that CH_2O-induced cross-links are removable by excision repair, and that this process can increase the frequency of single-strand breaks in the DNA of cells exposed to nonlethal concentrations of CH_2O.[14,18] The possible importance of cross-linking as a mechanism of initiation of CH_2O-induced nasal cancer in rats was pointed out by Swenberg et al.[15]

It would be expected from the foregoing that CH_2O would induce cross-links in the rat nasal mucosa *in vitro*. Evidence of cross-linking was obtained following brief reaction of CH_2O (2 min., 0°C, pH 8.0) with homogenates of the nasal mucosa of F-344 rats.[6] The reactions were followed immediately by reduction with $NaBH_4$, and the homogenates were solubilized using 8 M urea, 0.24 M phosphate, 10 mM EDTA, and 1% sodium dodecylsulfate (pH 6.8). The solutions were extracted with chloroform/*iso*-amyl alcohol/phenol (24/1/25), and the DNA was purified and analyzed by hydroxyapatite chromatography.

The amount of DNA that remained in the aqueous phase after extraction decreased with increasing concentrations of CH_2O

TABLE 3

Quantities of DNA in the Aqueous Phase after Incubation of Nasal Mucosal Homogenates with Formaldehyde and Extraction of the Homogenates with Chloroform/iso-Amyl Alcohol/Phenol (24/1/25)*

Formaldehyde Concentration** (mM)	Tissue Weight (mg)	DNA yield (μg)
0	80.8	225
3	73.8	248
30	82.0	83
300	74.0	10

* Homogenates were prepared from the anterior (respiratory) mucosa of sixteen rats (four per concentration) previously unexposed to CH_2O. The homogenates were maintained at 0°C in 0.1 M sodium phosphate, 5 mM EDTA (pH 8.0).

**Reactions were initiated by the addition of CH_2O followed 2 min. later by $[^3H]NaBH_4$ (37.5μCi; specific activity 5.0 μCi/mmole). Incubation was then continued for 30 min. The tissue was solubilized and extracted as indicated in footnote * of Table 2.

(Table 3). At higher CH_2O concentrations, a considerable amount of the DNA was not extracted into the aqueous phase, and was found to be in association with proteins at the interface between the aqueous and organic phases. This DNA was solubilized and recovered nearly quantitiatively by digestion of the interfacial material with proteinase K[6].

The nonextractability of DNA from proteins except after proteolysis has been cited previously as evidence of DNA-protein cross-linking, after treatment of cells with difunctional alkylating agents such as nitrogen mustards,[19-21] with CH_2O,[13] and with ultraviolet irradiation.[22] Although it is obvious that nonextractability of DNA is indirect evidence of cross-linking, other kinds of experiments, such as DNA alkaline elution,[14,15] cesium chloride density gradient untracentrifugation,[7,8] and gel electrophoresis[9,10] indicate that this interpretation is probably correct.

To further define the characteristics of nonextractable DNA in the rat nasal mucosa, homogenates were reacted with CH_2O *in vitro*, then they were solubilized and analyzed by isopycnic centrifugation in CsCl density gradient. Similar experiments were also performed on the nasal mucosa of rats exposed to 15 ppm of CH_2O *in vivo* (6 hr. daily for 2 days). The DNA of CH_2O-treated nasal mucosal samples banded at virtually the same density (approximately 1.68 g/cm^3), and the bands exhibited essentially the same width and shape as the DNA from control homogenates. Using isolated chromatin, it has been shown that cross-linking with proteins can cause a substantial shift of the DNA band to lower densities.[7,8] However, in those experiments the protein/DNA weight ratio and the concentration of cross-links (1 per 4 nucleotides in one case)[7] were both very high.

The concentration of cross-links in the present experiments was estimated using $^{14}CH_2O$. Under conditions that rendered 48% of the DNA nonextractable from proteins, there was less than 1 cross-link produced per 28,000 nucleotides. Calculations indicate that at this concentration of cross-links only a minor perturbation would result in the bouyant density of the DNA, consistent with the experimental finding.

The above observations suggest that the measurement of nonextractable DNA may be an extremely sensitive method for detection of DNA-protein cross-links *in vivo*. To examine whether CH_2O might induce the formation of cross-links in the nasal mucosa of F-344 rats, groups of animals were exposed to 15 ppm of CH_2O (6 hr. daily for 2 days), the nasal mucosa was dissected as soon as possible after exposure, and the tissue was homogenized and extracted in the same manner as was used previously for *in vitro* experiments. These experiments revealed a significant increase in the quantity of nonextractable DNA in the nasal mucosa of rats exposed to 15 ppm of CH_2O (Table 4)[6].

These results indicate the feasibility of detecting DNA changes *in vivo* by analysis of nonextractable DNA. However, because the yield of the latter may depend not only on the exposure regimen but also on other experimental parameters, careful adherence to a defined protocol is necessary to obtain reproducible results. Moreover, the relationship between the quantity of nonextractable DNA and the concentration of presumed cross-links in the DNA has not yet been determined.

In conclusion, we wish to stress that an increase in the quantity of nonextractable DNA is not in itself proof of cross-linking. Additional evidence using other methods is essential to place the present observations in a proper perspective. The significance of these observations to the carcinogenic mechanism of CH_2O is presently unknown. However, it is obvious that mechanistic studies cannot proceed without a

TABLE 4

Quantities of Nasal Mucosal DNA from Unexposed Male F-344 Rats Exposed to CH_2O (15 ppm, 6 hr. daily for 2 days) Recovered in the Aqueous Phase and in the Aqueous-Organic Interface after Extraction with Chloform/iso-Amyl Alcohol/Phenol (24/1/25)*

Location of DNA after Extraction	Average DNA yield (μg/mg wet wt. of tissue)**		Significance Level⁺
	Control	Exposed	
Aqueous	3.18 ± 0.43	2.84 ± 0.48	NS ($p > 0.25$)
Interface	0.49 ± 0.09	0.99 ± 0.16	Sig ($0.01 > p$)
(Total)	3.66 ± 0.44	3.82 ± 0.58	NS ($p > 0.5$)

* Respiratory mucosa was homogenized and incubated for 30 min. in the presence of [^{3}H]aBH$_4$ (7.5 mM; 5.0 μCi/mmole). Incubation conditions, solubilization, and extraction are indicated in the first footnote of Table 2.

**Values shown are means ± S.D. for five groups of control and exposed rats containing three animals per group.

⁺ One-tailed *t*-test; NS = Not Significant, Sig = significant.

clear understanding of the possible reactions that CH_2O can undergo with macromolecules in the nasal mucosa.

REFERENCES

1. D. French and J. T. Edsall, The reactions of formaldehyde with amino acids and proteins, Adv. Protein Chem., 2, 277 (1945).

2. M. Ya. Feldman, Reactions of nucleic acids and nucleoproteins with formaldehyde, Prog. Nucleic Acid Res. Mol. Biol., 13, 1 (1973).

3. L. Uotila and M. Koivusalo, Formaldehyde dehydrogenase from human liver. Purification, properties, and evidence for the formation of glutathione thiol esters by the enzyme, J. Biol. Chem., 249, 7653 (1974).

4. H. d'A. Heck, E. L. White and M. Casanova-Schmitz, Determination of formaldehyde in biological tissues by gas

chromatography/mass spectrometry, Biomed. Mass Spectrom., 9, 347 (1982).

5. M. Casanova-Schmitz and H. d'A Heck, Metabolism of formaldehyde in the rat nasal mucosa *in vivo*, The Toxicologist, 3, 61 (1983).

6. H. d'A. Heck and M. Casanova-Schmitz, Covalent binding of formaldehyde with macromolecules in the rat nasal mucosa, The Toxicologist, 3 (1983), In Press.

7. D. Brutlag, C. Schlehuber and J. Bonner, Properties of formaldehyde-treated nucleohistone, Biochemistry, 8, 3214 (1969).

8. D. Doenecke, Digestion of chromosomal proteins in formaldehyde-treated chromatin, Hoppe-Seyler's Z. Physio. Chem., 359, 1343 (1978).

9. Y. Ohba, Y. Morimitsu and A. Watarai, Reaction of formaldehyde with calf-thymus nucleohistone, Eur. J. Biochem., 100, 285 (1979).

10. V. Jackson, Studies on histone organization in the nucleosome using formaldehyde as a reversible cross-linking agent, Cell, 15, 945 (1978).

11. Y. F. M. Chaw, L. E. Crane, P. Lange and R. Shapiro, Isolation and identification of cross-links from formaldehyde-treated nucleic acids, Biochemistry, 19, 5525 (1980).

12. R. J. Wilkins and H. D. Macleod, Formaldehyde induced DNA-protein cross-links in *Escherichia coli*, Mutat. Res., 36, 11 (1976).

13. N. Magana-Schwenke and B. Ekert, Biochemical analysis of damage induced in yeast by formaldehyde. II. Induction of cross-links between DNA and protein, Mutat. Res., 51, 11 (1978).

14. W. E. Ross and N. Shipley, Relationship between DNA damage and survival in formaldehyde-treated mouse cells, Mutat. Res., 79, 277 (1980).

15. J. A. Swenberg, E. A. Gross, J. Martin and J. A. Popp, in "Proceedings of the Third CIIT Conference on Toxicology: Formaldehyde Toxicity," J. E. Gibson, ed., Hemisphere, Washington, D.C., 1982 (In Press).

16. C. Auerbach, M. Moutschen-Dahmen and H. Moutschen, Genetic and cytogentical effects of formaldehyde and related compounds, Mutat. Res., 39, 317 (1977).

17. D. L. Ragan and C. J. Boreiko, Initiation of C3H/10T1/2 cell transformation by formaldehyde, Cancer Lett. 13, 325 (1981).

18. A. J. Fornace, Jr., Detection of DNA single-strand breaks produced during the repair of damage by DNA-protein cross-linking agents, Cancer Res., 42, 145 (1982).

19. R. J. Rutman, W. J. Steele and C. C. Price, Experimental chemotherapy studies. II. The reactions of chloroquine mustard and nitrogen mustard with Ehrlich cells, Cancer Res., 21, 1134 (1961).

20. R. H. Golder, G. Martin-Guzman, J. Jones, N. O. Goldstein, S. Rotenberg and R. J. Rutman, Experimental chemotherapy studies. III. Properties of DNA from Ascites cells treated *in vivo* with nitrogen mustard, Cancer Res., 24, 964 (1964).

21. O. Klatt, J. S. Stehlin, Jr., C. McBride and A. C. Griffin, The effect of nitrogen mustard treatment on the deoxyribonucleic acid of sensitive and resistant Ehrlich tumor cells, Cancer Res., 29, (1969).

22. P. Todd and A. Han, in "Aging, Carcinogenesis, and Radiation Biology," K. C. smith, ed., Plenum, New York, 1976, p. 83.

DISCUSSION

MR. MONAGHAN: (Leo Monaghan, Monsanto.) The rate of cross-linking should be a function of the pH. Did you measure the pH in this system at all?

DR. HECK: *In vitro* we did all our reactions at pH 8. *In vivo*, of course, the pH would be the normal physiological pH, which is about 7.4.

MR. MONAGHAN: Just looking at the chemistry of the system as a polymer chemist, not a biochemist, I would expect these cross links to occur primarily on the acid side.

DR. SWENBERG: (James Swenberg, CIIT.) Dr. Heck, just to embellish this a little bit, there was a paper at the Cancer Meeting this last year in which work from Al Fornace's lab showed cross-linking using alkaline elution with human tracheal explants, and all of the cross-links were repaired within a

3-hour period afterwards. So, it does appear that at least human tracheal cells have a rapid repair system for DNA cross-links. We had shown and reported at the previous Formaldehyde Conference that V79 cells have a similar type of activity.

DR. HECK: Yes, the cross-links are definitely not permanent. They can be repaired by enzymatic methods. It is well known that that occurs.

DR. BERNSTEIN: (Martin Bernstein, Ciba-Geigy.) What if the mucus was stripped out of your preparation? If you pretreated the animal with something that dried up the mucus first, and then ran this experiment, would that lead to more binding? In other words, are you perhaps just reacting with a protein coat whose removal could lead to more cellular binding?

DR. HECK: To repeat, we are measuring DNA-protein cross-linking, and certainly the protein or the mucous coat effects could be very important in that. We have not done the low concentration experiments yet, so I cannot tell you what the results will be as we go down in concentration.

CHAPTER 10

THE EFFECT OF FORMALDEHYDE EXPOSURE ON CYTOTOXICITY AND CELL PROLIFERATION

J. A. Swenberg, E. A. Gross
H. W. Randall and C. S. Barrow

Chemical Industry Institute of Toxicology
Research Triangle Park, North Carolina

Animals exposed to toxic concentrations of formaldehyde gas develop mild to severe lesions in the epithelium lining the nasal cavity. Such toxic injury induces an increase in cell proliferation in the respiratory epithelium. This compensatory response replaces dead and dying cells as well and increases the thickness of the epithelium. Since cell proliferation is a sensitive indicator of toxicity and is required for both initiation and promotion of chemical carcinogenesis, we have investigated the effect of formaldehyde exposure on cell turnover. The data indicate that an initial wave of cell replication occurs in rats and mice 18 hours after a 6 hr. exposure to 15 ppm. The percent of replicating cells remains markedly elevated for 3-5 days and then begins to decrease. Similar elevations occurred following 3 daily exposures to 6 ppm formaldehyde in rats, but not mice. No increase occurred in rats or mice exposed to 0.5 or 2 ppm. Additional studies on rats exposed to a daily Concentration x Time product of 36 ppm-hrs. have demonstrated that increased cell proliferation correlates with concentration, not C x T.

INTRODUCTION

After discovering that exposure of rats to high concentrations of formaldehyde caused squamous cell carcinomas of the nasal passages[1,2], we began a series of studies to understand the pathogenesis of this lesion. Initially our research centered on defining the acute toxic response. Cytotoxicity of nasoturbinate was evident in 1 μ plastic-embedded sections after a single-6 hour exposure to 15 ppm formaldehyde[3]. Following additional exposures, a series of cellular changes in the nasoturbinate were apparent, ranging from mild rhinitis to severe ulceration. After as little as 5 days' exposure to 15 ppm formaldehyde, species differences between rats and mice were striking[4]. One hundred percent of the rats had ulceration of the nasoturbinate, while only focal cellular degeneration was evident in mice. The primary mechanism thought to be responsible for this marked species difference is an adaptive response by mice which decreases their minute volume, thereby resulting in less formaldehyde being deposited in the nasal passages[4,5]. Furthermore, when rats and mice are chronically exposed to 15 ppm formaldehyde, the theoretical dose deposited in the nasal passages of mice is about half that of rats (Fig. 1). Thus, the nasal passages of mice exposed to 15 ppm formaldehyde are subjected to toxic insult similar to that of rats exposed to 6 ppm. This calculated similarity correlates well with the final incidence of squamous cell carcinomas in our chronic toxicity study[2], where the adjusted incidence was 1% in rats exposed to 6 ppm versus 3.3% in mice exposed to 15 ppm (Table 1).

Cytotoxicity and Cell Proliferation Following Formaldehyde Exposure

A prominent result of formaldehyde's cytotoxicity was restorative cell proliferation and hyperplasia. Increased numbers

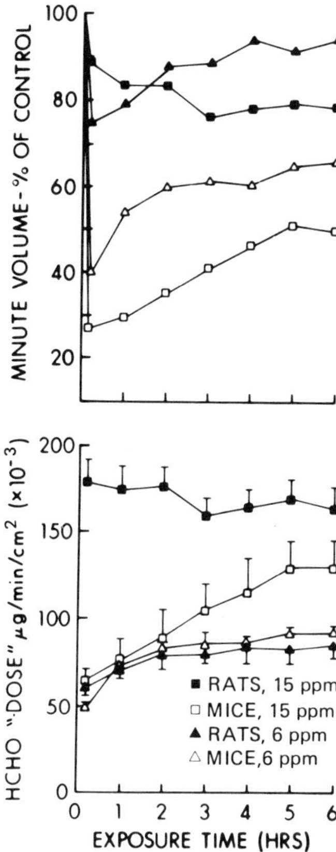

Fig. 1. Effect of formaldehyde exposure on minute volume and theoretical "dose" of rats and mice pretreated with the same exposure for 4 days. Data from Reference 4.

TABLE 1

Adjusted Incidence of Squamous Cell Carcinoma of the Nasal Passages Following Inhalation Exposure to Formaldehyde for 24 Months

Formaldehyde Concentration (ppm)	Number of Tumors/Animals at Risk: Rat	Number of Tumors/Animals at Risk: Mouse
0	0/208 (0%)	0/72 (0%)
2	0/210 (0%)	0/64 (0%)
6	2/210 (1%)	0/73 (0%)
15	103/206 (50%)	2/60 (3.3%)

of mitotic figures and thickening of the nasal epithelium were evident following 3-9 days exposure to formaldehyde. By administering a pulse of ^{3}H-thymidine (2μCi/g) to animals immediately following formaldehyde exposure and preparing light microscopic autoradiographs of the nasal passages, we were able to quantitate the extent of cell proliferation in rats and mice exposed for various periods of time to different concentrations of formaldehyde.

METHODS AND RESULTS

Pilot studies indicated that maximal cell proliferation occurred after 3-5 days of formaldehyde exposure and that proliferation decreased after 9-10 days[3]. Using the three-day time point, we investigated the concentration response of cell proliferation in rats and mice. Animals were labeled with ^{3}H-thymidine 2 hours after the third day of exposure to either 0, 0.5, 2, 6 or 15 ppm formaldehyde. Level B of the nasal turbinates was selected for autoradiography (Fig. 2). This is an area that is extensively cleansed by the respiratory mucociliary apparatus, compared to Level A[6]. Table 2 shows that rats exposed to 6 or 15 ppm had a 10-20 fold increase in the number of labeled respiratory epithelial cells, compared to controls, whereas only mice exposed to 15 ppm formaldehyde exhibited a similar increase. No increase in cell proliferation relative to control animals was evident in rats exposed to 0.5 or 2 ppm or in mice exposed to 0.5, 2 or 6 ppm.

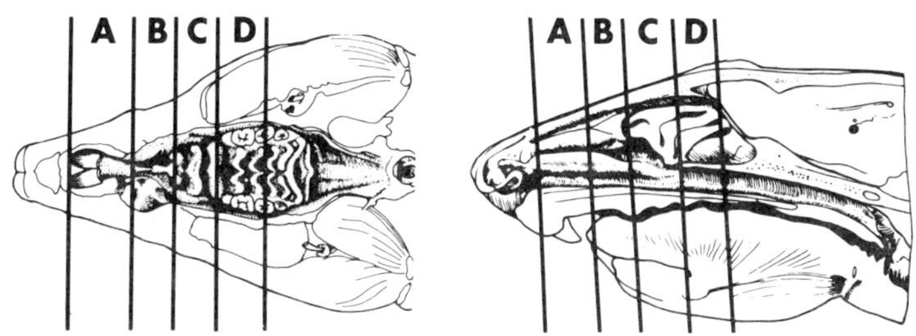

Fig. 2. Drawing indicating the level of sections from the nasal passages of rats and mice.

TABLE 2

Effect of Formaldehyde Exposure on Cell Proliferation in Level B of the Nasal Passages

Exposure*	% of Labeled Respiratory Epithelial Cells**	
	Rat	Mouse
Control	0.22 ± 0.03	0.12 ± 0.02
0.5 ppm	0.38 ± 0.05	0.09 ± 0.04
2 ppm	0.33 ± 0.06	0.08 ± 0.04
6 ppm	5.40 ± 0.82	0.15 ± 0.06
15 ppm	2.83 ± 0.81	0.97 ± 0.04

* All animals exposed for 6 hrs./day for 3 days.
**Mean ± standard error.

It was unclear, however, whether or not the pulse of ^{3}H-thymidine 2 hours after exposure represented the most sensitive time for increased cell labeling, since many toxicants and carcinogens can cause an initial inhibition of DNA synthesis. To investigate this, rats were exposed for 3 days to 6 ppm formaldehyde. Animals received a pulse of ^{3}H-thymidine 2 or 18 hours after the last 6-hour exposure. The results are shown in Table 3. The percent of labeled cells doubled when animals were pulsed 18 hours after exposure. Control animals exhibited a similar increase, presumably due to circadian variations. Whether or not this difference results in greater sensitivity, will be tested by a companion study to the data in Table 2 that is currently underway.

Additional investigations demonstrated that as little as a single 6-hour exposure to 15 ppm formaldehyde caused a tenfold increase in cell proliferation in both rats and mice when the animals were pulsed 18 hours post-exposure (Table 4). Examination of the autoradiograms from the 1-day-exposed animals revealed that nearly all of the nasal cells were labeled. Thus, a 6-hour exposure to 15 ppm formaldehyde is an adequate stimulus to synchronize the respiratory epithelium of the nasal passages to undergo a marked proliferative response. This labeling index further increased to over 20 times that of controls when rats were exposed to 15 ppm for 5 days and labeled 18 hours later. In

TABLE 3

Effect of the Time of ^{3}H-Thymidine Pulse on Cell Replication after Formaldehyde Exposure to Rat

Post-Exposure Time of Pulse	% Labeled Cells*	
	0 ppm	6 ppm**
2 hour	0.26 ± 0.03	1.22 ± 0.17
18 hour	0.54 ± 0.06	3.07 ± 1.09

* Mean ± standard error.
**6 ppm, 6 hr./day for three days.

TABLE 4

Effect of Single or Repeated Formaldehyde Exposure (15 ppm, 6 hr./day) on Cell Proliferation in Level B Respiratory Epithelium*

Exposure	F-344 Rats	B6C3F1 Mice
Control	0.43 ± 0.05**	0.27 ± 0.04
1 day	5.51 ± 0.35	2.14 ± 0.56
5 day	10.05 ± 0.27	3.42 ± 0.84

* All animals were pulsed with ^{3}H-thymidine (2 μCi/gm) 18 hours after the last exposure.
**Mean ± standard error.

contrast, only a small additional increase in labeling index was noted for mice.

Effect of Concentration Versus Cumulative Exposure

Previous comparisons of 6-month exposure data demonstrated striking differences in the pathological response to 15 ppm formaldehyde, 6 hrs./day, 5 days/week (450 ppm-hrs./week)[2,7] vs. 3 ppm, 22 hrs./day, 7 days/week (462 ppm-hrs./week)[8]; with rats from the latter study exhibiting much less toxicity[2]. A study was therefore designed to evaluate the impact of concentration versus cumulative exposure on cytotoxicity. Preliminary data clearly indicate that concentration plays a major role in cytotoxicity and cell proliferation. Table 5 shows the experimental design, whereby all formaldehyde-exposed animals received the same daily cumulative exposure, i.e., 36 ppm-hours. A strong correlation was present in Level B of the rat nasal passage between concentration and cell proliferation, with a

TABLE 5

Effect of Formaldehyde Concentration vs. Cumulative Exposure on Cell Turnover in Rats (Level B)

Exposure	% Labeled Cells*	
	3 Days + 18 Hr.	10 Days + 18 Hr.
Control	0.54 ± 0.03	0.26 ± 0.02
3 ppm x 12 hr.	1.73 ± 0.63	0.49 ± 0.19
6 ppm x 6 hr.	3.07 ± 1.09	0.53 ± 0.20
12 ppm x 3 hr.	9.00 ± 0.88	1.73 ± 0.65

* Mean ± standard error.

marked increase over controls after 3 days exposure to 12 ppm, versus an intermediate effect at 6 ppm and a small increase at 3 ppm. The labeling index decreased after 10 days' exposure in all groups, with sevenfold increase at 12 ppm versus a doubling at 3 and 6 ppm. This correlates with the histology, since the hyperplastic state had been attained by 10 days' exposure. Thus, the 3-day response represented a combination of hyperplasia due to thickening of the epithelium and compensatory proliferation for cell death, while the 10-day response was primarily composed of the latter.

In Level A, where mucociliary clearance is minimal, the C x T product appears to be more important than concentration. Table 6 demonstrates a fivefold increase in the labeling index of all formaldehyde-exposed rats after 3 days' exposure. Note also the increase in the labeling index of control cells in the anterior nose. The only location of Level A that exhibited a concentration effect was the medium septum.

TABLE 6

Effect of Formaldehyde Concentration vs. Cumulative Exposure on Cell Turnover in Rats (Level A)

Exposure	% Labeled Cells After 3 Days Exposure*
Control	3.00 ± 1.56
3 ppm x 12 hr.	16.99 ± 1.50
6 ppm x 6 hr.	15.46 ± 10.01
12 ppm x 3 hr.	16.49 ± 2.07

* Mean ± standard error.

In the same C x T study, mice presented a much different response. There was no increase in labeling index or any evidence of cytotoxicity in Level B following 10 days of exposure. In Level A a trend was apparent in which cell proliferation was inversely proportional to concentration after 10 days (Table 7). Since data are not yet available for 3-day exposures in mice, caution must be exercised when interpreting these data. Additionally, due to minute volume differences between species which affect C x T products, the cumulative dose is not the same. This is evident in Table 8 where mice exposed to 3 ppm for 12 hours are predicted to receive over 50% more formaldehyde each day versus rats from those C x T groups or rats and mice of the other two C x T groups. This may help explain why mice were more severely effected at the 3 ppm x 12 hour exposure. It does not explain, however, why lesions in rats extended more posteriorly (to Level B) than in mice at their respective C x T exposure groups. The differences observed may be

TABLE 7

Effect of Formaldehyde Concentration vs. Cumulative Exposure on Cell Turnover in Mice (Level A)

Exposure	% Labeled Cells After 10 Days Exposure*
Control	1.24 ± 0.57
3 ppm x 12 hr.	10.14 ± 3.20
6 ppm x 6 hr.	4.72 ± 1.61
12 ppm x 3 hr.	1.76 ± 0.49

* Mean ± standard error.

due to other factors such as higher linear velocity of the airstream through the nose of rats and/or the presence of better mucociliary flow and cellular repair mechanisms in mice. This apparent species difference, in view of similar "dose" predictions, requires further study.

TABLE 8

Prediction of Theoretical "Dose" to the Nasal Cavity in Rats and Mice Exposed to Formaldehye in C x T Study*

Exposure (C x T)	"Dose"/Minute (μg/min/cm^2)		"Dose"/Day (μg/day/cm^2)	
	Rats	Mice	Rats	Mice
3 ppm x 12 hrs.	0.036	0.059	25.9	42.5
6 ppm x 6 hrs.	0.083	0.083	29.9	29.9
12 ppm x 3 hrs.	0.143	0.148	25.7	26.6

* "Dose" prediction derived from data reported in References 4 and 5.

DISCUSSION

From these studies, it is clear that cell proliferation is concentration, dose- and time-dependent. Species differences in upper respiratory tract reflexes during formaldehyde exposure exert considerable influence on the extent of the response. In rats, where a similar minute volume is breathed over formaldehyde exposures of 2-15 ppm, concentration is the major contributor to the extent of toxicity and cell turnover in respiratory epithelium. In the mouse, the concentration effect is less clear because different doses occur even though C x T products of exposure are the same. It is also evident that the greatest response occurs in the most anterior respiratory epithelium, that region of the nasal passages having the least protection from mucociliary clearance. Lastly, no increase in cell proliferation was detectable at 0.5 or 2 ppm formaldehyde. Some caution must be exercised in this conclusion, however, since pulse labeling was done only at 2 hours after exposure.

ACKNOWLEDGMENTS

The authors would like to thank Messrs. W. Steinhagen and M. Phelps for providing the animal exposures. Drs. L. Morgan and J. Chang are thanked for their assistance and intellectual contributions.

REFERENCES

1. J. A. Swenberg, W. D. Kerns, R. I. Mitchell, E. J. Gralla and K. L. Pavkov, Induction of squamous cell carcinomas of the rat nasal cavity by inhalation exposure to formaldehyde vapor, Cancer Res., 40, 3398 (1980).

2. Chemical Industry Institute of Toxicology/Battelle Columbus Laboratories, Final Report, Docket #10922. A chronic inhalation toxicology study in rats and mice exposed to formaldehyde, CIIT, Research Triangle Park, NC, 1981, 4 vols.

3. J. A. Swenberg, E. A. Gross, J. Martin and J. A. Popp, Mechanisms of formaldehyde toxicity, in "Proceedings of the 3rd CIIT Conference on Toxicology: Formaldehyde Toxicity," J. E. Gibson, ed., Hemisphere, Washington, D.C. (In Press).

4. J. C. Chang, E. A. Gross, J. A. Swenberg, and C. S. Barrow, Nasal cavity deposition, histopathology and cell proliferation following single or repeated formaldehyde exposures in B6C3F1 mice and F-344 rats, Toxicol. Appl. Pharmacol. (In Press).

5. J. C. Chang, W. H. Steinhagen, and C. S. Barrow, Effect of single or repeated formaldehyde exposure on minute volume of B6C3F1 mice and F-344 rats, Toxicol. Appl. Pharmacol., 61, 451 (1981).

6. K. T. Morgan, D. L. Patterson, and E. A. Gross, Formaldehyde and the mucociliary apparatus, in "Formaldehyde: Toxicology, Epidemiology and Mechnanisms," J. J. Clary, J. E. Gibson and R. S. Waritz, eds., Marcel Dekker, New York, 1983, p. 193.

7. W. D. Kerns, D. J. Donofrio, and K. L. Pavkov, The chronic effects of formaldehyde inhalation in rats and mice. A preliminary report, in "Proceedings of the 3rd CIIT Conference on Toxicology: Formaldehyde Toxicity," J. E. Gibson, ed., Hemisphere, Washington, D.C. (In Press).

8. G. M. Rusch, Sub-chronic and chronic continuous exposure, multi-species studies with formaldehyde, in "Proceedings of the 3rd CIIT Conference on Toxicology: Formaldehyde Toxicity," J. E. Gibson, ed., Hemisphere, Washington, D.C., (In Press).

CHAPTER 11

MECHANISMS OF FORMALDEHYDE TOXICITY AND RISK EVALUATION

T. B. Starr

Chemical Industry Institute of Toxicology
Research Triangle Park, North Carolina

Recent quantitative risk assessments of formaldehyde have used data from the CIIT-sponsored two-year chronic formaldehyde inhalation study with Fischer 344 rats. The multi-stage model employed for low-dose extrapolation had the concentration of formaldehyde in inspired air, i.e., the administered dose, as its independent variable. Implicit in the use of this exposure measure is the assumption of strict linear proportionality between administered dose and the amount of formaldehyde that reaches target cells, i.e., delivered dose. At high doses, however, the structure and function of the rat nasal mucosa are qualitatively different from what they are at low doses. For example, mucostasis, ciliastasis, enhanced cell replication, and physiological compensation occur at high doses, but apparently not at low doses. Each of these changes can alter the ratio of delivered to administered dose, producing a non-linear and species-specific relationship between these two dose measures. The implications of such non-linearities in the mucosal delivery system are illustrated by contrasting quantitative risk assessments that are identical except in the use of administered or delivered dose as the independent variable.

INTRODUCTION

With few exceptions, current regulatory practices impose an important constraint on the quantification of risk from exposure to potential human carcinogens. This constraint requires that numerical estimates of risk be derived exclusively from the tumor incidence data provided by chronic whole animal bioassays. Other information on the toxicity of a chemical compound, regarding, for example, its mutagenicity in *in vitro* test systems, or from detailed mechanistic studies of the compound's deposition, distribution, metabolism, and pharmacokinetics, is not currently factored into numerical estimates of risk. Instead, such information is used selectively only to defend the choice of a particular mathematical model for performing low dose extrapolations. This is a serious issue because the wealth of mechanistic data that is available on formaldehyde is having no direct impact on numerical risk estimates. What seems to be needed is a way to make maximal scientific use of this kind of mechanistic information in the process of risk quantification.

One vehicle for doing this is the concept of delivered dose. This term denotes the amount of the final active form of a compound that is present at a specific target site. Delivered dose should be contrasted with administered dose, the traditional measure of exposure in chronic bioassays, expressed, for example, as the concentration of a compound in ambient air for exposures by inhalation.

Quantitative risk assessments based exclusively on bioassay data assume implicitly that the relationship between administered and delivered doses is linear across the entire dose spectrum from infinitesimally small to nearly lethal doses. If this linear proportionality assumption is valid, then the need for mechanistic data is completely obviated, because administered dose is then a perfect proxy measure for delivered dose. However, if this assumption is invalid, the relationship between administered and delivered doses can be almost anything, and bioassay data alone, which are essentially descriptive in nature, can tell us nothing about it.

Mechanistic information is required to test the validity of the linear proportionality assumption. Further, should linear proportionality prove to be invalid, mechanistic information is absolutely necessary to specify just what the non-linear relationship is between these two dose measures. Given this relationship, bioassay data can be analyzed with delivered dose as the independent variable. Since this dose measure reflects the number of molecules of the active form of a compound that actually reach a target site, quantitative risk estimates based on delivered dose will presumably be more realistic and scientifically defensible than those based solely on administered dose.

It will be shown in what follows that numerical estimates of risk based on delivered and administered doses can differ dramatically from one another when the relationship between these two dose measures is non-linear, even if the same mathematical model is used to represent the dose-response relationship in both cases. For vinyl chloride, a compound that requires metabolic activation in order to produce toxicity, use of delivered dose leads to more conservative risk estimates that are consistent with all of the bioassay data[1].

Formaldehyde, in contrast, requires no metabolic activation. However, it may be removed from proximity to the target site by normal mucociliary clearance processes or detoxified at the target site by normal metabolic processes that serve to regulate endogenous formaldehyde levels. When these protective responses are modeled with Michaelis-Menten kinetics similar to those used to describe the metabolic activation of vinyl chloride, the resulting non-linear relationship between administered and delivered doses leads to risk estimates that are significantly lower than corresponding ones based exclusively on administered dose. In order to place these risk estimates in perspective, I first present a review of various procedures for producing quantitative risk estimates exlusively from the CIIT sponsored chronic formaldehyde inhalation toxicity study.

METHODS, RESULTS, DISCUSSION

Risk Estimates Based on Administered Dose

Table 1 presents the ambient air concentrations of formaldehyde at which selected analysis methods estimate the risk of developing squamous cell carcinomas in the nasal cavity to be 1/100,000 for Fischer 344 rats exposed 6 hours per day, 5 days per week, for 24 months. Incidence data were taken from the CIIT-sponsored chronic inhalation toxicity study. The number of rats having squamous cell carcinomas/number sacrificed at 24 months was 0/160, 0/160, 2/160 and 87/160 at 0.0, 2.0, 5.6, and

TABLE 1

Ambient Air Concentrations of Formaldehyde at which the Risk to F-344 Rats under the CIIT Chronic Bioassay Exposure Protocol is Estimated to be 1/100,000.

Method		C (PPM) for Risk = 10^{-5}
Linear Extrapolation		
From 14.3 ppm:	VSD	0.00026
	LCL95	0.00024
From 5.6 ppm:	VSD	0.0045
	LCL95	0.0018
NOEL with Safety Factor		0.002
Multi-Stage Models		
2-stage	VSD	0.060
	LCL95	0.006
Epigenetic w/Safety Factor		0.006
3-stage	VSD	0.342
	LCL95	0.006

14.3 ppm formaldehyde in ambient air, respectively. These incidence data yield the most conservative proportions possible apart from those which would also include animals with certain benign tumors of an entirely different cell type in origin.

The linear extrapolation method (LE) simply involves connecting the proportion responding at a particular administered dose, or its upper 95% confidence limit, to the origin with a straight line. This method is widely recognized as being the most conservative of those currently in use. The 14.3 ppm dose group was the only one to exhibit a response that was significantly different from that in the controls ($p < 0.05$). Using the response in this group, the LE method yields a best estimate (lower 95% confidence limit (LCL95) of 0.26 ppb (0.24 ppb) for the ambient air concentration at which the risk is

1/100,000. Using the response in the 5.6 ppm dose, it yields a best estimate (LCL95) of 4.5 ppb (1.8 ppb).

To put these estimates in perspective, the average daily concentration of formaldehyde in the ambient air of Los Angeles in 1979 has been reported to be 15 ppb[3]. Thus, if no strain differences exist in the sensitivity of rats to the effects of formaldehyde, assuming that the ambient air concentration does not deviate markedly from the reported daily average, and further assuming that the LE estimate based on the 14.3 ppm response is correct, one would expect at least 57/100,000 Los Angeles rats 24 months or older to exhibit squamous cell carcinomas of the nasal cavity upon necropsy.

The traditional toxicological approach to risk assessment for non-carcinogenic endpoints involves the determination of a No-Observed-Effect-Level (NOEL). This level is then adjusted downward by a Safety Factor which is presumed to account for any variability in sensitivity within and across species. If this procedure is used with 2.0 ppm as the NOEL and with a conservative Safety Factor of 1000, the concentration at which the risk is estimated to be 1/100,000 is 2.0 ppb, nearly the same as the LE LCL95 based on the response at 5.6 ppm. It should be noted that this method fails to take into account the steepness of the dose-response. Thus, the result would be the same if the first positive response occurred at 560 ppm rather than at 5.6 ppm.

Several quantal response models have been proposed for quantitative risk assessment[4]. Of these, the multi-stage model seems to be the most widely accepted, since initiation and promotion have been identified as steps in the tumor induction process for genotoxic carcinogens, and since the multi-stage model includes a two-stage model as a special case. It should be noted, however, that the biological plausibility of the multi-stage model is contingent on its use with a dose measure that is at the very least a valid proxy for delivered dose. Thus, if the relationship between delivered and administered doses is non-linear, the multi-stage model need not be biologically plausible with administered dose as its independent variable.

Table 1 presents maximum likelihood estimates of the virtually safe doses (VSD's) and their LCL95's for two- and three-stage models fit to the formaldehyde incidence data. Although the three-stage VSD exceeds the two-stage VSD by nearly a factor of five, note that the LCL95's for both models are the same, 0.006 ppm. This is due to the presence in both LCL95 models of a term that is linear in administered dose, while only a quadratic (cubic) term is present in the maximum likelihood form of the two-stage (three-stage) model.

Recently, a proposal has been put forth for a modified quantitative risk assessment procedure to deal with epigenetic carcinogens (compounds that cannot initiate by themselves but can promote spontaneously otherwise initiated cells). It consists of using the maximum likelihood form of the multi-stage to estimate the administered dose at which 10% incidence is expected. This dose is then scaled with a safety factor as in the traditional toxicological approach for non-carcinogenic compounds. Such a procedure presumably accomodates the existence of a threshold exposure level for promotion effects. For the two-stage model, 10% incidence is predicted to occur at 6 ppm formaldehyde, so with a safety factor of 1000, the resulting safe exposure level is 0.006 ppm, and the same as the LCL95's for the two- and three-stage models. The steepness of the dose-response for formaldehyde exposure thus leads to the same low-exposure levels irrespective of whether or not the analysis procedure assumes formaldehyde genotoxicity.

As noted earlier, other quantal response models can be fit to the formaldehyde incidence data. In addition to linear extrapolation and the multi-stage, Fig. 1 illustrates the relationship between maximum likelihood risk estimates and administered dose for the Weibull, multi-hit, logit and probit models. For risks on the order of 1/100,000, the VSD's are all within an order of magnitude of each other, the exception being the linear extrapolation result. However, as the level of risk decreases, the spread between the model estimates increases as indicated. It should also be noted that if additive background contributions to risk are allowed or forced to enter, all of the models become linear at sufficiently low doses, and the risk estimates end up parallel to the linear extrapolation curve.

Figure 2 displays the incidence data and the two-stage model that was fit to them as a function of administered dose. The model overestimates the response at 5.6 ppm by nearly an order of magnitude and also noticeably underestimates the response at 14.3 ppm. The steepness of the observed dose-response is the source

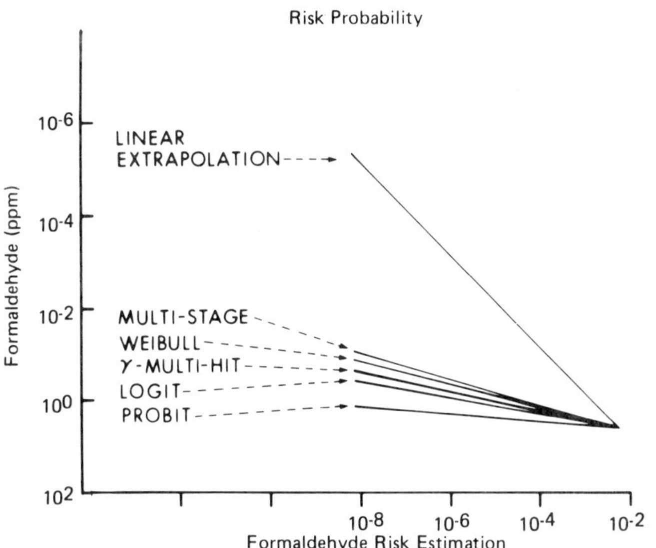

Fig. 1. Estimated risk versus ambient air concentration of formaldehyde for linear extrapolation and selected quantal response models.

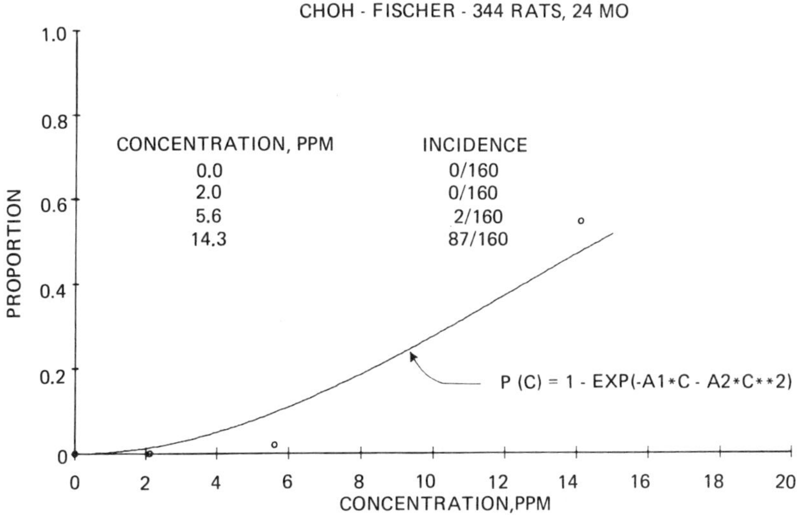

Fig. 2. Incidence of F-344 rats having squamous cell carcinomas of the nasal cavity at 24 months and the fitted two-stage model incidence.

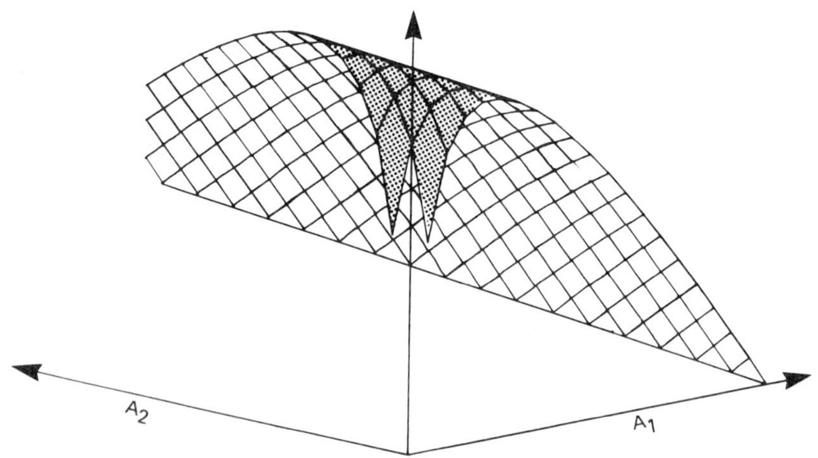

Fig. 3. Log-likelihood surface for the 2-stage model as a function of the linear and quadratic parameters A1 and A2.

of this difficulty. A three or higher stage model would be required to provide a more satisfactory fit.

Figure 3 displays the log-likelihood surface for two-stage models of the formaldehyde bioassay data. The independent variables are the two parameters A1 and A2 of the two-stage model equation displayed in Fig. 2. (An independent background incidence parameter, A0, was taken to be identically zero since every model that allowed for independent background incidence has a maximum likelihood estimate (MLE) of it which was zero.)

MLE procedures seek that combination of the independent variables that corresponds to the highest (least negative) point on the log-likelihood surface. For two-stage models, the search is restricted to the first quadrant (A1 and A2 both $>$ 0), since other combinations result in the prediction of negative incidence at some doses when the background incidence is zero. It is evident from Fig. 3 that the restricted maximum occurs for A1 = 0. However, this is not a global maximum because higher values of the log-likelihood occur in the second quadrant (A1 $<$ 0, A2 $>$ 0).

It is also evident from Fig. 3 that the 90% confidence region surrounding the restricted maximum will invariably include non-zero values of A1. This feature is responsible for the introduction of a linear term in the LCL95 of multi-stage VSD's which is the dominant one for sufficiently low levels of risk.

Risk Estimates Based on Delivered Dose

As noted earlier, mechanistic data provide the information necessary to establish the validity or invalidity of the linear proportionality assumption that underlies administered-dose-based risk estimates. Given the ubiquity of non-linearity in the world of biology, it is somewhat surprising that this extraordinary assumption has been so readily and generally accepted without solid scientific evidence to support it. However, it must be recognized that replacement of this assumption with a non-linear relationship between administered and delivered doses requires a tremendous amount of mechanistic information on compound-specific deposition, disposition, metabolism, and pharmacokinetics in whole animals. Such information is rarely available. One compound for which it is available is vinyl chloride.

Vinyl Chloride

Fig. 4 illustrates the administered dose-response relationship for the incidence, at death, of hepatic angiosarcomas in

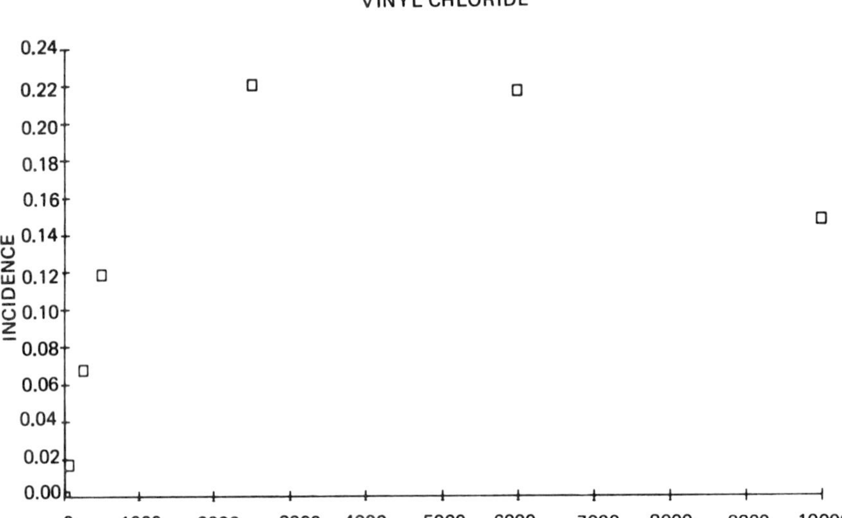

Fig. 4. Incidence of Sprague-Dawley rats having hepatic angiosarcomas at death following exposure to vinyl chloride by inhalation for 4 hours/day, 5 days/week, for 52 weeks.

Sprague-Dawley rats exposed to vinyl chloride by inhalation for 4 hours/day, 5 days/week, for 52 weeks[5]. This dose-response is unusual in that it is steepest at the lowest administered doses. Among multi-stage models, the one-stage model provides the best fit. However, the quality of the fit is poor, and the estimated risk at low doses appears to be anti-conservative due to the influence of the plateau in incidence at high administered doses on the one-stage parameter[6]. Similarly anti-conservative low dose risk estimates would be generated by the linear extrapolation procedure described earlier.

This problem has been resolved by modeling the incidence data with delivered dose as the independent variable[7]. Vinyl chloride must be metabolically activated before it can produce hepatic angiosarcomas, and the biotransformation of vinyl chloride in rats proceeds in accordance with Michaelis-Menten, i.e., saturable, kinetics[8]. Consequently, there is an upper limit on the rate of production of the active metabolite responsible for tumor induction. Thus, although linear proportionality between administered and delivered doses can be expected to hold at low administered doses, it fails at sufficiently high administered doses.

This departure from linearity is illustrated in Fig. 5 using the non-linear relationship reported by Gehring _et al_[9]. If the hepatic angiosarcoma incidence data are plotted against the delivered doses produced by biotransformation of the administered doses as in Fig. 6, a much 'better behaved' dose-response results. It is nearly linear in the low-dose region, so the estimated risk associated with low doses using the multi-stage model would be consistent with all of the data, in contrast with the apparently anti-conservative estimates based on the administered dose-response of Fig. 4.

Formaldehyde

Turning now to formaldehyde exposure by inhalation, we can inquire whether similar considerations regarding the delivery of formaldehyde to target tissue can have a bearing on the estimation of low-dose risk using the bioassay incidence data. The target tissue in this case is presumed to be the basal cells capable of replication that underly respiratory epithelium in the rat nasal cavity.

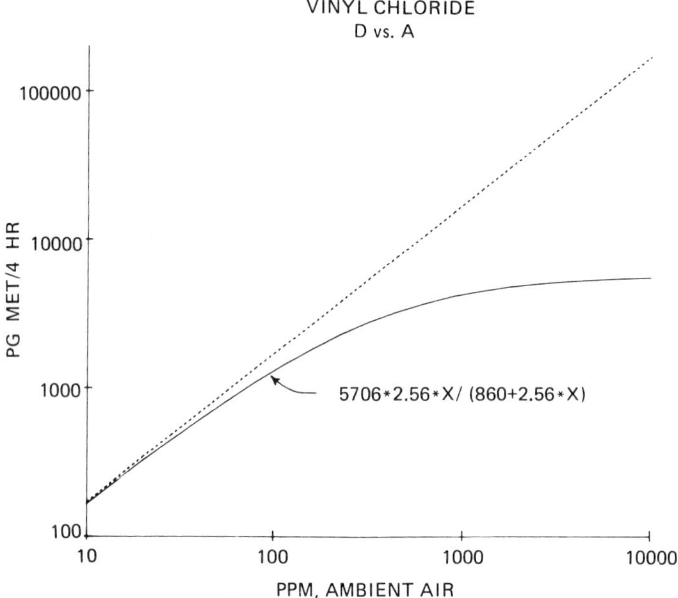

Fig. 5. Delivered dose versus administered dose for vinyl chloride.

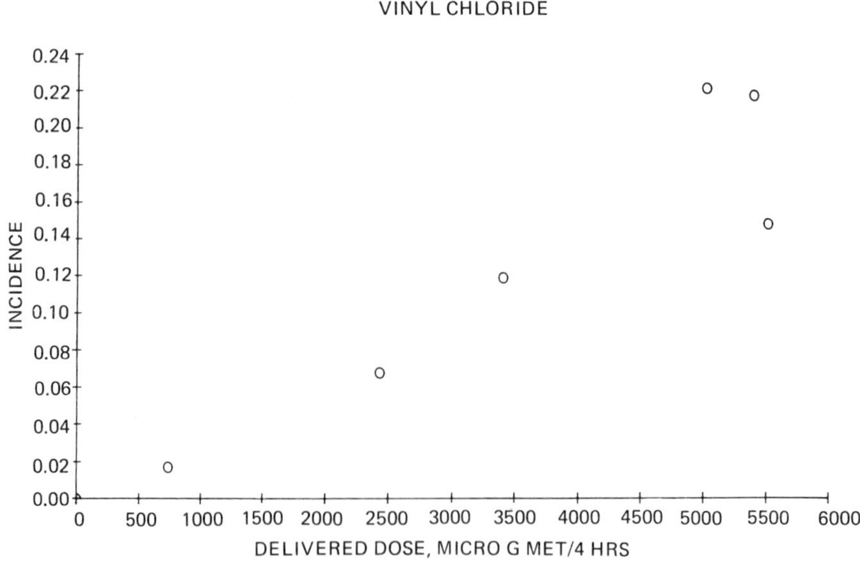

Fig. 6. Same as Fig. 4 except abscissa is the delivered dose of Fig. 5.

In order to form a permanent lesion on the DNA of such cells, inhaled formaldehyde must first be deposited on the outer surface of the mucous blanket. This dynamic protective layer consists primarily of water and extended glycoproteins. It serves to warm and humidify inhaled air and also to trap particulate matter before it reaches the lungs. It can also serve as a sink for inhaled formaldehyde, since the glycoproteins provide sites for binding in protein cross-links. At low ambient air concentrations, formaldehyde deposited on the mucous blanket also can be removed from proximity to the target tissue via normal mucociliary clearance processes.

Should formaldehyde penetrate the mucous blanket, it must then diffuse through the underlying periciliary fluid, escape removal via intercellular fluids, including the blood, and then penetrate the basal cell membrane. Once inside the basal cells, it must escape the normal metabolic detoxification processes that serve to regulate endogenous formaldehyde levels. Finally, formaldehyde must penetrate the nuclear membrane, bind to DNA, and this lesion must escape DNA repair processes until the occurrence of cell division.

It should be apparent from the above remarks that the delivery of formaldehyde to specific target cells consists of a complicated series of processes even when exposure is by inhalation and the target tissue is cells of the respiratory epithelium in the nasal cavity. There are a number of routes by which inhaled formaldehyde can be removed from the target site, and also by which damage to DNA can be eliminated should it occur. Consequently, it is reasonable to presume that the amount of formaldehyde fixed in DNA lesions that escape repair can be calculated as the difference between the amount deposited on the surface of the mucous blanket and the amount removed, detoxified, or otherwise accomodated by the respiratory epithelium of the nasal cavity. Furthermore, it is reasonable to presume that these routes of removal and repair are governed by saturable kinetics similar in kind to those responsible for the metabolic activation of vinyl chloride.

A hypothetical relationship between administered and delivered doses for the case of formldehyde is presented in Fig. 7. This relationship embodies the above consideration by assuming that the rate of delivery is the difference between the rate of deposition and the rate of removal and repair, the latter being governed by Michaelis-Menten kinetics with a saturation constant of 1 ppm formaldehye in ambient air, and with the rate of removal and repair equal to the deposition rate in the low-dose limit.

In the observable response range of administered doses, the relationship is nearly linear, since the removal and repair processes are saturated in this range. Below 1 ppm, however, the relationship departs significantly from linearity, and is essentially quadratic below about 0.05 ppm. Indeed, at 0.006 ppm, the LCL95 for the two- and three-stage models at which the risk was estimated to be 1/100,000, the delivered dose predicted by this relationship is approximately 0.000036 ppm. Now the estimated risk for the two- and three-stage models is linear in dose

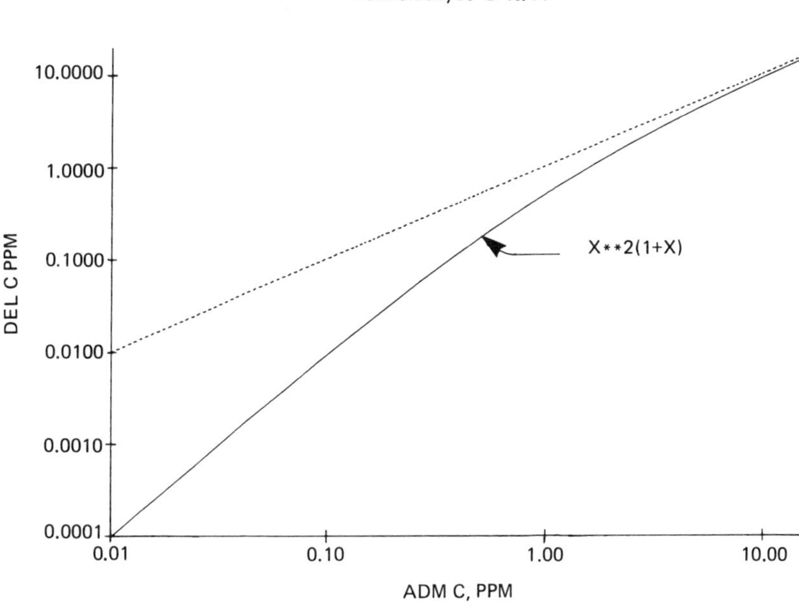

Fig. 7. An hypothetical non-linear relationship between delivered and administered doses of formaldehyde assuming Michaelis-Menten type kinetics for removal, metabolic detoxification and repair.

for these small administered doses. It follows that the risk estimate of 1/100,000 based on administered dose is nearly 167 times higher than that based on the corresponding delivered dose given by this relationship.

Alternatively, one can calculate the administered dose required to produce an estimated risk of 1/100,000 based on 0.006 ppm delivered. This yields 0.081 ppm administered, over 13 times larger than the 0.006 ppm predicted with administered dose. Thus, a mild Michaelis-Menten type non-linearity can have a dramatic impact on the estimates of risk associated with low ambient air concentrations of formaldehyde.

CONCLUDING REMARKS

It could be argued, of course, that there is as yet no direct evidence that delivered dose is non-linearly related to administered dose below the observable response range. This is certainly true. However, there is direct evidence that normal mucociliary clearance is rapidly and significantly inhibited for administered doses of 15 ppm[10]. Further, marked qualitative changes in the structure and function of the rat nasal mucosa, including severe ulceration of the nasoturbinates, have been reported after as little as 5 days of exposure to 15 ppm formaldehyde[11]. There is thus strong direct evidence that the removal and repair processes which could protect basal cells from formaldehyde induced damage at low administered doses are themselves severely disrupted in the observable response range. Of course, verification of their protective role at low ambient air concentrations and identification of the precise nature of the non-linear relationship between administered and delivered doses require further mechanistically oriented research.

In summary, there should be no question that the delivered dose concept can have a dramatic impact on numerical estimates of risk. Vinyl chloride provides an excellent and well-documented example. A most important issue which remains to be resolved is whether the delivered dose concept is required for the case of formaldehyde exposure by inhalation. Although the answer is not currently unequivocal, the preliminary and indirect evidence points toward the affirmative.

REFERENCES

1. H. M. Bolt, H. Kappus, R. Kaufmann, K. E. Appel, A. Buchter and W. Bolt, "Metabolism of 14C-vinyl chloride _in vitro_ and _in vivo_," Colloq. Inst. Natl. Sante Rech. Med., 52, 151 (1975).

2. Chemical Industry Institute of Toxicology/Battelle Columbus Laboratories, "Final Report. A chronic inhalation toxicology study in rats and mice exposed to formaldehyde," CIIT, Research Triangle Park, NC, 1981, 4 vols.

3. Versar, Inc., "Draft Report (Contract No. 68-01-5791)," for Office of Pesticides and Toxic Substances, U.S. Environmental Protection Agency, Springfield, VA, 1980, pp. 70, 103, 107.

4. D. Krewski and J. Van Ryzin, "Dose response models for quantal response toxicity data," in "Statistics and Related Topics," M. Csorgo, D. A. Dawson, J. N. K. Rao and A. K. M. dE. Saleh, eds., North-Holland, New York, 1981, p. 201.

5. C. Maltoni and G. Lefemine, "Carcinogenicity assays of vinyl chloride: Current results," Ann. N.Y. Acad. Sci., 246, 195, (1975).

6. J. Van Ryzin and K. Rai, in "The Scientific Basis of Toxicity Assessment", H. Witschi, ed., Elsevier/North-Holland, Amsterdam, 1980, p. 273.

7. R. H. Reitz, J. F. Quast, A. M. Schumann, P. G. Watanabe and P. H. Gehring, "Non-Linear Pharmacokinetic Parameters Need to be Considered in High Dose-Low Dose Extrapolation," In "Quantitative Aspects of Risk Assessment in Chemical Carcinogenesis," Arch. Toxicol., Suppl. 3, 79 (1980).

8. P. G. Watanabe, R. E. Hefner, Jr. and P. J. Gehring, "Fate of (14C) vinyl chloride following inhalation exposure in rats," Toxicol. Appl. Pharmacol., 37, 49 (1976).

9. P. J. Gehring, P. G. Watanabe and C. N. Park, "Resolution of dose-response toxicity data for chemicals requiring metabolic activation: example vinyl chloride," Toxicol. Appl. Pharmacol., 44, 581 (1978).

10. K. T. Morgan, "Formaldeyde and the Mucociliary Apparatus," in "Formaldehyde: Toxicology, Epidemiology and Mechanisms," J. J. Clary, J. E. Gibson and R. S. Waritz, eds., Marcel Dekker, New York, 1983, p. 193.

11. J. C. Chang, E. A. Gross, J. A. Swenberg and C. S. Barrow, "Nasal cavity deposition, histopathology and cell proliferation following single or repeated formaldehyde exposures in B6C3F1 mice and F-344 rats," Toxicol. Appl. Pharmacol. (In Press).

DISCUSSION

DR. CROCKER: (Timothy Crocker, University of California.) You gave a very fine figure for the dose delivered in micrograms per minute per square centimeter of nasal mucosa. What was not clear to me was how you calculated that. Would you care to run through that?

DR. STARR: I did not calculate that number, but the methodology is to make a measurement of the entire surface area of the nasal cavity. One assumes that all of the formaldehyde that is inhaled during the course of 1 minute of inhalation is deposited within the nasal cavity. You take that number of micrograms of formaldehyde, divide it by the surface area of the nasal cavity and express it as a dose of micrograms per minute per square centimeter of surface area.

MS. MARGOSCHES: (Elizabeth Margosches, Environmental Protection Agency.) What are your feelings about the fact that a great many of the mathematical models come from an essentially similar family, regardless of whether you are using administered dose or delivered dose? Also, what are your thoughts about the fact that the safe doses for several of them clustered and clustered some distance from the linear model?

DR. STARR: I would be hard pressed to argue that low-dose linearity should not apply, if the independent variable is delivered dose and we are dealing with a genotoxic carcinogen. We don't know this is the case with respect of formaldehyde. Even though formaldehyde has been demonstrated to be mutagenic/genotoxic in test systems of one kind or another, we don't know that it is in the human case. We don't even know that it is in the rat case. If it were an inactivated genotoxic mechanism in the rat, that does not guarantee that it is going to be an epigenetic mechanism in the human. One possible explanation for the steepness of this dose-response curve is that it is, in fact, just promoting spontaneous tumors and that you need enhanced cytotoxicity in order for that effect to result. I don't understand your question with respect to them all coming from the same family. I don't think they necessarily do.

MS. MARGOSCHES: Several of them do.

DR. STARR: If anything, I would expect that if one were to have the right dose measure, the disparity between virtually safe doses predicted by those different models would disappear.

MR. DU PONT: (Norman DuPont, Paul, Hastings.) What does it take before you determine that delivered dose is the best or the only method to be used in assessing risk for formaldehyde?

DR. STARR: This is opinion, but it seems to me that all you have to do is demonstrate unequivocally that there is a qualitive difference between the structure and the function of the system that is involved in the delivery at high doses as contrasted with low doses. We are not trying to predict human risk from exposure to 15 parts per million. We want to do it at 1/10 part per million, something like that, and if there is a difference, then you will have to reflect on that difference and consider seriously whether a non-linear relationship between delivery and administration would be required.

With respect to formaldehyde, there is literally wholesale destruction -- not just saturation, of whatever protective mechanisms are present at low doses. So, in this case, I guess my opinion is that we have demonstrated that consideration of delivered dose is necessary; what we have not provided yet is the specific recipe. Further work is going to be needed to do that.

CHAPTER 12

CONFERENCE SUMMARY AND PERSPECTIVES

DR. CLARY: I would now like Dr. James Gibson from CIIT, who is the Vice President and Director of Research, to both lead us in general discussion and do the wrap-up. Because of schedule conflicts, Dr. Harris will not be able to do the latter. Dr. Gibson will set his own order for the discussion/ wrap-up.

DR. GIBSON: (James Gibson, CIIT.) Many, if not all of the speakers are still here, so your unanswered questions can be discussed. Before we go into questions from the floor, I would like to take John's invitation literally and attempt to sum up. I would like to do it using one figure (Fig. 1).

The figure is one that I used in the CIIT Conference on formaldehyde two years ago. It was used in a presentation where I attempted to sum up the present knowledge of the mechanism of

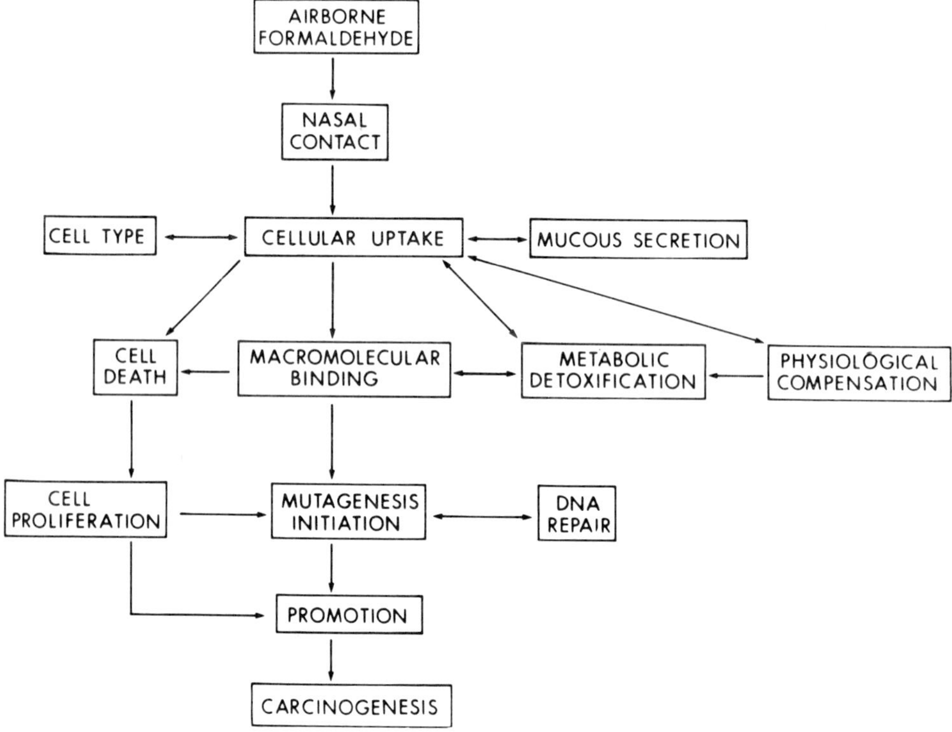

Fig. 1. Possible metabolic pathways for exogenous formaldehyde. (Reproduced by permission of Hemisphere Publishing Co., New York.)

formaldehyde's carcinogenesis. I proposed at that time that the process was almost certainly a multi-stage process. Many of the boxes that were presented raised interesting questions for which we had no answers.

As a result of work accomplished over the past two years, many of the questions that were raised have now been answered. Of course, that is not to say that there are no remaining questions.

The basic problem that confronted us at that time, and now, is the question of how airborne formaldehyde leads to cancer in the nose of rats and mice.

We proposed the possibility that one explanation could follow a straight path mechanistically from the presence of formaldehyde in the air to cancer. In fact, if we follow that straight path from the top of the figure to the bottom and that way of thinking, we have the fairly standard dogma of chemical carcinogenesis. But there are some alternative paths which we proposed two years ago for which I think today's proceedings have possibly provided some additional support.

In the early work we were concerned about the contact of airborne formaldehyde with the nasal cavity and whether or not formaldehyde would be taken up by cells. Of course, this is not an issue at all. Nor is the question of whether formaldehyde causes cancer in rats and mice, since we all agree that it does.

The question is the mechanism. Two years ago we wondered about the modifying influence of mucous secretions on cellular uptake of formaldehyde. At that time we had no information at all. Morever, we wondered if there was some influence of cell type in the nasal mucosa; one cell type that would be more susceptible than another. I don't believe that question has yet been answered, but it may be an important one. Certainly the variety of cell types found in rats, mice, and people could and should be examined for common denominators.

We do know now, as a result of Dr. Morgan's work, that mucus may very well influence the cellular uptake of formaldehyde. Especially at low concentrations of formaldehyde in air, the mucous blanket may function very effectively as a removal process.

We know that cell death can result from formaldehyde that does reach the cells. This has been known from the early work. But the studies that we heard from Dr. Swenberg have elaborated on that in more detail. Indeed, he has demonstrated that as a result of cell death, cell proliferation likely is initiated. The question of whether formaldehyde penetrates the various cellular barriers to reach genetic material, which is what concerns us when we think of macromolecular binding, has also been addressed. A number of studies have demonstrated the potential of formaldehyde to cause genetic toxicity.

Biochemically, interactions with DNA have been measured; yet the question of the importance of this in *in vivo* systems remains to be further elucidated. We also know that formaldehyde that is taken up by the cells is readily metabolized. The metabolizing system for formaldehyde in cells 1) is extremely effective; 2) does not seem to be saturated at high concentrations in air; and 3) continues to function during chronic exposure to formaldehyde. Thus, detoxification continues to occur as does physiological compensation as we have heard concerning the difference between rats and mice.

The question of macromolecular binding leading to mutagenesis, I have already mentioned. Certainly this is a possibility, and these mutational events may serve to initiate the process of carcinogenesis in specific cells.

The question of DNA repair is open. We have heard of some studies where DNA repair mechanisms for formaldehyde-protein-DNA cross-links are readily repaired. But this area needs additional work.

Cell proliferation in itself is of interest in this regard when the number of cells dividing is greatly increased. The probability of an error during replication can creep in and may be a mechanism for initiation quite independent of macro-molecular binding.

Finally, initiated cells can be promoted with formaldehyde by the simple process of cell proliferation or irritation, and carcinogenesis can result.

The figure from two years ago continues to be as good now as it was then, I think, although we do know more about mucous secretion. We do know much more about cell proliferation, and I think we do know much, much more about the importance of expressing the dose in terms of delivered dose rather than administered dose. It is not sufficient to take the amount of airborne formaldehyde as the independent variable, as Dr. Starr has told us.

Two years ago in the CIIT Conference, in addition to papers on toxicity studies and mechanistic studies, we heard papers on some epidemiology work. More information on this important subject has been added today and it is reassuring to note that the study at Du Pont and the study in the Ontario morticians have not discovered any increased incidence of cancers among those exposed people. Of course, we are awaiting with great interest the results of the large National Cancer Institute/Formaldehyde Institute epidemiology study.

Today, we heard more about the question of initiation and promotion for formaldehyde, which we suggested sometime ago would be of considerable interest. These skin painting studies also provided some reassuring information that formaldehyde may, under the conditions of the conducted experiments, be only a weak promoter.

Additional information on the potential mutagenicity of formaldehyde has been added by the observations reported on the TK locus in human cells by Dr. Goldmacher.

The work of CIIT concerning mucociliary clearance, biochemical reaction, cell turnover mechanisms and low-dose extrapolation methodology is important. We are proud of it, and additional work will continue.

So, in the past two years, considerable progress has been made. The primary contribution has been the realization that for formaldehyde, an important endogenous biochemical, the delivered dose must be used in any low-dose extrapolation method that we undertake and probably it is less important exactly what that method happens to be.

With that brief overview of the proceedings today, contrasted against where we were two years ago, I would like to invite the audience to pose questions to the speakers. I would also be happy to address any questions.

DISCUSSION

DR. STERNBERG: (Steven Sternberg, Memorial Cancer Center, Sloan Kettering.) I would like to know what Dr. Starr meant when he said that formaldehyde does not have to be activated. Is that because it is such a simple compound? Can he explain that a little further to a non-biochemist?

DR. GIBSON: I suggest Dr. Heck address that question, since he has been researching that subject. I think there is a fairly direct answer for it.

DR. HECK: (Henry Heck, CIIT.) By activation one usually means conversion to a product that is more reactive than the starting material. In the case of formaldehyde, of course, the product of metabolism is not reactive. It is just formate, and so formaldehyde itself is activated, as it were, already.

DR. BERNSTEIN: (Martin Bernstein, Ciba-Geigy.) Dr. Gibson, given all this new data, is the mouse a better model for formaldehyde risk extrapolation to humans than is the rat?

DR. GIBSON: No. I don't think that is what any of the data says at all. Dr. Swenberg and Dr. Starr both went to some length to explain that the responses in mice and rats seem to be equivalent when we calculate the delivered dose for those two species. So, there isn't any reason to believe that there is any innate sensitivity of one species over the other. So, for this particular agent, rat and mouse may both be appropriate species. What the data do say is that when we do these experiments and we interpret them, we had better pay attention to the amount of the active material that is reaching the target site, rather than just the amount that is in the air.

MR. MONAGHAN: (Leo Monaghan, Monsanto.) One of the uses of knowledge of the mechanism is to develop a way of preventing this toxic effect from occurring. Do you have any means, any additives or anything at all that might adequately interfere with the proposed mechanism so that the mouse or the rat can tolerate higher concentrations of formaldehyde?

DR. GIBSON: I think perhaps we could develop some interference, but I don't know why we would. In this particular instance what all the work shows, as I have repeatedly said, is that the really important consideration is the amount of stuff that gets there. If we can prevent that amount from reaching the site or keep that amount at sufficiently low concentration so that no qualitative changes occur in the physiological function of the animal, that might be the best way. But you could conceive of ways of increasing mucous secretion or mucus flow and so forth and, based on what we heard from Dr. Morgan, perhaps attentuate the formaldehyde effects somewhat.

DR. MC MARTIN: (McMartin, LSU Medical Center.) In my view, formate is not necessarily considered a non-toxic metabolite of formaldehyde. I was wondering if any of the people that did the initiation and promotion studies have ever done, or know of any studies done, on the effects of formic acid as an initiator or promoter?

DR. GIBSON: Dr. Boreiko at CIIT has conducted studies using the 10T1/2 mouse fibroblast intitiation, a neoplastic transformation assay, to assess formaldehyde effects. Simultaneously with that, he has studied formic acid. He finds in that system that formaldehyde can behave as both an initiator and promoter of neoplastic transformations, but only at very high concentrations. Formaldehyde does not act as a complete carcinogen in that system. At best, its activity is considered weak, and formic acid is negative in that system for initiation, promotion, and complete carcinogenicity.

DR. RUSCH: (George Rusch, Allied Chemical.) I have two questions for Dr. Krivanek regarding the skin painting studies. The first regards mixing the formaldehyde in the acetone: Has anyone checked to see whether there could be a reaction of the formaldehyde with the acetone to decrease the actual concentration of the material being applied? The second is in two parts and concerns the two higher levels of formaldehyde that were applied in the Haskell Laboratory study. Both appeared to be irritating levels and I am looking at the inverted dose-response curve. First of all, was an irritation seen during the actual administration at these levels, and second, could the irritation have caused a decrease in the actual dose due to a sloughing of the epithelial tissue? I'm looking for some more insight into why that dose-response curve appears to be inverted.

DR. KRIVANEK: The question of a reaction occurring between formaldehyde and acetone has not been checked. We do know, when we did the preliminary studies, that we did get skin responses with the mixture, so we know that something is occurring. Of course, this gets to the question of the liver dose and what was the actual dose that the animals were getting. There are, also, further complications on any kind of study like this. One of them is how much of the formaldehyde has evaporated into the air due to the skin temperature and the air flow in the test chamber or in the hood when the material is applied. All these things do occur, and we have not gone to the lengths required to get answers on these questions. Regardless, this is the result of this study. Now, conceivably, one could go to much higher concentrations of formaldehyde and really give the skin a workout, perhaps using long-term continuous application, and produce ulcerations. That, we think, would be confounding the results of the experiment. So that has not been done. But these things all could be looked at.

Regarding the inverse dose-response: Again, the possibility exists that, at the very low concentration we did not get sloughing of skin cells so that the material stayed there longer and was able to exert a stronger effect. That presently is not being studied. Again, as I pointed out in the presentation, one doesn't know if one is looking at biological variation or if one is looking at this question of sloughing of the cells.

When we did treat the animals, we found that initially there is what you might call drying of skin. It is just visually observed. Afterwards, the animals, I hate to use the word "hardened," but that seems to be what happened. All these things could happen, and certainly studies like this generally lead to other studies where one has to look at how much is getting into the skin and getting through the skin to act with the cells and produce an effect. So these are model systems. The initiation-promotion type of model is more of a screening procedure.

DR. RUSCH: Do you know if there were any apparent differences in the skin tone for the three dose levels that were used in the 26-week application?

DR. KRIVANEK: No, there were not. As I said initially, there is some drying at the higher dose. In a week or two the animals again appeared normal. At the end of the study the skin looked quite like normal mouse skin at 52 weeks of age. There was essentially no difference.

DR. KANG: (Han Kang, OSHA.) I have a question for Dr. Swenberg. Have you attempted to estimate or calculate the delivered dose in human nasal turbinates?

DR. SWENBERG: No. We don't have any comparable surface area data. Our surface area data were derived by doing serial sections of the entire nasal cavity and using computerized morphometry. There is no comparable data available on the human nose at this time. It could be done, and it should be done.

DR. KANG: Eventually, the delivered dose depends on the minute volumes, does it not?

DR. SWENBERG: Yes, that is part of delivered dose.

DR. KANG: And the surface area.

DR. SWENBERG: Also, the mucus flow and all the other points that were in Dr. Starr's figure. It is not just the amount that is breathed in that comprises delivered dose. It is the amount that gets to the DNA that comprises delivered dose.

DR. KANG: No. My understanding of your calculation for rats and mice was that the delivered dose had a twofold difference.

DR. SWENBERG: Don't confuse dose that we had in quotes with Dr. Starr's delivered dose. Those are two different numbers. One is a calculated dose to the animal. That has taken care of the normalization across species that Dr. Starr mentioned. So we have done the first cut in his delivered dose concept, but you have many additional steps to look at.

DR. KANG: I understand that, but somebody tried to explain the incidence of a major carcinoma based on that delivered dose. The species difference can be explained by the amount delivered to the major surface area.

DR. SWENBERG: Incidentally, a major part of it.

DR. KANG: My question is: What is the difference in minute volume between say, rat or mouse, and man? Was that ratio determined?

DR. SWENBERG: There are data available on minute volumes. I am not an expert in that area. I cannot give you just what the numbers are, as far as human minute volume. There are no comparable data on surface area, so we cannot get the human equivalent at this time. It is attainable if we get proper morphometry of the human nose.

DR. KANG: I have a question for Dr. Levine. Can you please compare you mortician's exposure level to occupational exposure?

How did your mortician exposure level compare to a worker who is exposed to 1 ppm for 40 hours a week?

DR. LEVINE: I think Dr. Kang wants me to bring out the fact that the morticians are exposed intermittently to low levels of formaldehyde. Furthermore, not all of them are exposed during every year that they were observed in the study because some do not have any work that will involve formaldehyde exposure and some do. I think, also, in that situation we are covering a large number of years, going back to 1928. I really don't know the levels of formaldehyde exposure likely to have occurred in those years. Perhaps others in the audience might be able to give us a better idea about that. It is my understanding that with the passage of time, the concentrations of formaldehyde in embalming fluids have gradually been reduced as, for example, more effective delivery of the formaldehyde to the tissues has been achieved through use of penetrating chemicals and so forth. Thus, in times past, we might have had greater exposures, and so forth. Possibly, also, embalmers might have mixed their own chemicals way back in time. I cannot say. However, at least at present day, the exposures for embalmers would be quite low. As I mentioned, it would be about 0.3 parts per million for the embalming of a normal body, which comprises the majority of the bodies that an embalmer would embalm. There might be on the order of a couple of bodies a week, for a total of several hours exposure at that low level for those people who actually embalm. I don't know if I can relate that to a worker who is exposed 40 hours a week and so forth. Also, the exposed worker may not be exposed 40 hours a week, either. The worker may be exposed when, for example, opening the door of an oven or going by the open oven but not be exposed during other parts of his job. So his exposure, as well, may be intermittent. If you wanted me to calculate it in terms equivalent to a 40-hour-per-week exposure, we might say that maybe today the average embalmer is getting the equivalent of a couple of weeks of 40-hour exposures at something less than 1 part per million over a year's time.

MS. PERERA: (Frederica Perera, NRDC.) This is really related to Dr. Kang's question. Is there evidence that humans can physiologically compensate for formaldehyde exposure by decreasing minute volume as has been shown for the mice?

DR. GIBSON: The effect that we observed in the mice that seems to be most important, is the one at 15 parts per million and perhaps to a lesser extent at 6 parts per million. While rats have this respiratory reflex at high concentrations, it doesn't seem to affect them as severely as the mice. Consequently the mice respond more significantly. As these animals are exposed to smaller concentrations, then this sensory irritation reflex disappears and it becomes of virtually no importance in modifying the amount of formaldehyde being breathed.

I think it is probably known that humans have these sensory irritation reflexes, and they may be very important at high concentrations. At low continuous concentrations they may not be important, the same as in mice and rats.

MS. PERERA: Have any studies been done with human exposure to formaldehyde?

DR. GIBSON: Yes, Dr. Ib Anderson has conducted such studies looking at a variety of responses. One of his interests, of course, was mucociliary clearance, and his studies were mentioned today.

MS. PERERA: I mean in terms of decrease in minute volume.

DR. GIBSON: I don't know of specific studies that have been done in that regard. Do you Dr. Swenberg?

DR. SWENBERG: No. But I can speak personally. When I inhale formaldehyde in my work as a pathologist, I hold my breath. I think humans are known to have the dive reflex, and that is essentially the same reflex that the mice are exhibiting here. So this becomes a factor when humans encounter high concentrations. But, as Dr. Gibson said, at concentrations, say, of half a part per million or lower, it is not a reflex that is brought into play at all. So, it doesn't become an important factor for these low dose conditions, only for the high dose.

MR. WANDS: (Ralph Wands, The Mitre Corporation.) There is a great deal of emphasis these days, in the regulatory community particularly, on structure-activity relationships. I would like to have your comments on what your results with formaldehyde, as have been discussed today, might mean in terms of other chemicals of similar structure? For example, acrolein, which is also a strong irritant and an aldehyde.

DR. GIBSON: The question is, given the results with formaldehyde, does this cause us to wonder about other aldehydes? The answer, of course, is yes. At CIIT we are planning some additional studies with other agents, including acrolein, but there are studies underway in other laboratories around the country and the world which are addressing these problems as well.

Dr. Feron in Holland is looking at acetaldehyde and finds that it, also, is irritating to the nasal cavity and can cause tumors there. But the tumors, as I understand it, are localized in the olfactory epithelium rather than in the respiratory epithelium. Thus, with agents having similar chemical structures or reactivity you might expect similar results. But the results, while they may be similar in some regards, can show some differences. When we compare formaldehyde's potency with acetaldehyde, I believe there is a difference of about two orders of magnitude in the dose that is needed to produce the effect. Acetaldehyde requires a much, much larger concentration than formaldehyde in air to produce irritation and tumors.

DR. MORGAN: (Kevin Morgan, CIIT.) I would just add something to that, if frogs are anything to go by: 1500 ppm of acetaldehyde gives a response in frogs very similar to 15 ppm of formaldehyde, i.e., stimulation and subsequent inhibition. The difference is exactly the two orders of magnitude in concentration.

DR. CROCKER: (Timothy Crocker, University of California.) Following up this question of the dose delivered as compared to the dose administered. I take it we really are saying that the mouse and rat are different, chiefly on the strength of the reduced minute ventilation that the mice underwent during that reflex reduction in tidal volume as a result of the irritant. But we really would like to know what fraction of the airborne concentration of formaldehyde is actually removed in a pass through the nose. Have you done anything to look at this?

DR. SWENBERG: (James Swenberg, CIIT.) There are some studies which show that formaldehyde is almost entirely removed by the nose. I cannot recall the authors. In addition to those studies, autoradiographs of animals who have received ^{14}C formaldehyde by inhalation in nose-only exposure for 30 minutes, do not show it getting beyond the nasal cavity and then the digestive tract.

DR. CROCKER: Those very often assume, or are based on, resting ventilation. But if we were to assume any changes in the ventilatory rate and the flow rate through the nose, we might then begin to have to challenge the assumption. We do all acknowledge that there is a major amount of absorption in the nose, that the assumption perhaps is correct and that all one needs is tidal volume. But there are flow rate questions, and I am wondering if you have approached them?

DR. SWENBERG: No, we haven't, and you are absolutely correct in that. Also, if you mouth-breathe, of course, you don't have the nasal absorption. It is an important point.

DR. STERNBERG: (Steven Sternberg, Memorial Cancer Center, Sloan Kettering.) I would like to remind everyone that when you hold your breath, the next breath after that is a deep one.

DR. CLARY: (John Clary, Celanese.) Well taken.

DR. MORGAN: (Kevin Morgan, CIIT.) One comment with respect to the difference between the rodent and man and mouth breathing. When we breathe through our mouth, some goes through our nose, and some goes through our mouth. It is a split stream, but the nasal epithelium still gets exposed. We are saying this is a very different situation, but I think we had better remember that whereas taking the rat respiratory epithelium as a model for the human respiratory epithelium, there appears to be no evidence for cancer induction in squamous epithelium in the rat. The cancer did not start at the front of the nose, which gets a high dose. If you want to start comparing the rat data with respect to risk of the pharynx, then I think you are confusing the issue because there it is squamous. It is not respiratory.

DR. CLARY: This concludes the meeting. I want to thank the authors of the papers. We have appreciated the papers very much and the audience for sticking with us. We will get the Proceedings out quickly. Have a safe journey home.

INDEX

A

Acetaldehyde, nasal irritation by, 274
Alcohol and cancer, 137-139
Angiosarcoma, hepatic, from vinyl chloride, 247, 249
Autoradiographs, 228, 231, 275

B

BAP [*see* benzo(a)pyrene]
Benzo(a)pyrene, 159, 161, 163, 166, 168

C

Carcinogen, alkylating, 190
Carcinogen:
  complete, 147, 151, 152, 156
  genotoxic, 256
Carcinogen, epigenetic, 243
Carcinogenesis:
  mechanism, 261
  multi-stage, 259
Carcinogenicity, formic acid, 269
Carcinoma, squamous cell, 32 168, 194, 226, 240, 242
Case-control study, 50
Cell proliferation, 160, 228, 230, 233-235, 237, 262, 263
Cell replication (*see* Cell proliferation)
Cell transformation, 160
CHAT medium, 175, 176
Chemical Industry Institute of Toxicology (*see* CIIT)
CIIT inhalation study on formaldehyde, 31, 32, 237, 240
Ciliastasis, 195, 197, 198, 201, 205, 206, 237
Cohort, 127
Cohort mortality study, 128, 129
Consumer Product Safety Commission, 33, 38, 43
Cumulative exposure index, 76

D

Deaths due to cancer, 54, 62, 64-67, 70, 73-75, 78-88, 116, 132
DNA:
  excision repair, 216
  nasal mucosa content, 216-18, 219
  rapid repair system, 223
  reaction with formaldehyde, 211, 213, 214, 251, 262
  repair, 251, 262
  single strand breaks, 216
  unscheduled synthesis, 174
Dimethylbenzanthracene, 147-150, 152, 153
DMBA (see dimethyl-benzanthracene)
Dose, administered, 237-239, 253
Dose, administered vs. delivered, 237, 238, 245, 247, 254, 263
Dose, delivered, 237-239, 247, 256, 257, 264, 265, 269
Dose-response curve:
  general, 34, 35
  of vinyl chloride, 247

E

Epidemiology studies, 24, 26, 49, 128, 263
Epithelium, respiratory, 235, 249
Equivalent lifetime contin-uous exposure, 33, 37

F

Formaldehyde:
  acetone, non-reaction with, 267
  air concentrations:
    garment manufacturing, 10, 16, 18
    hospitals, 10, 12, 18
    laboratories, 10, 12
    offices, 10, 19, 20
  analysis:
    area, 15, 18-20
    N-benzylethanolamine for, 4, 27
    3-benzyoxazolidine and, 5
    CEA instruments analyzer, 3, 14, 15, 17, 18
    charcoal tube method, 4, 6, 14-16, 18
    Chromosorb 102[R] for, 4, 6, 14, 15, 18, 28
    chromotropic acid for, 3
    diffusion badges for, 28, 29
    Draeger[R] detector tubes for, 2, 3, 14
    by Infra-red light absorption, 24
    NIOSH method P&CAM, 125, 4, 28
    _____ ______ _____, 318, 4
    _____ ______ _____, 354, 5
    personal, 14-16, 19, 29
    sodium bisulfite for, 14
  bound, 211-215
  cancer risk, 108 (*see also* Chapts. 2 & 11)
  carcinogenicity mechanism 259
  ciliastasis by, 206
  cytotoxicity of, 225
  effect on cell prolifera-tion, 225, 226, 228
  ______ __ nasal mucosa, 254
  ______ __ minute volume, 226
  embalming fluid concen-tration of, 271
  metabolism of, 211, 262, 262, 265, 266
  mutagenicity of, 173, 264

G

Genotoxicity, 43, 243, 256

H

Hyperplasia, cellular, 226, 232

I

Index, labelling, 232, 233
Industrial hygiene surveys:
  general, 7, 11, 21
  methodology, 2
Initiating agents (see Initiation, cancer)
Initiation, cancer, 147-157, 159, 163, 166, 169, 191, 225, 243, 263, 266

K

Keratocanthoma, 168

L

Low dose extrapolation:
  general, 237, 264
  models for:
    gamma (see multi-hit)
    linear, 31, 36, 42, 237, 241, 244, 256
    logic, 35, 244
    multi-hit, 35, 244
    multi-stage (see polynomial)
    polynominal, 36-38, 237, 242, 244, 246, 249, 252
    probit, 35, 244
    threshold, 38
  Weibull, 35, 36, 38-40, 42, 44, 45, 244
Lymphoblast TK6 cells, culture of, 175-177, 180
Lymphoma cells, mouse, 187, 188

M

Matched-pairs study, 50
Mechanisms:
  carcinogenesis, 261
  cell repair, 235
Mice, SENCAR, 147-151, 153-155
Minute volume, 270
Mitotic figures, 228
Mucociliary apparatus, 193, 205, 206, 211, 228
Mucociliary clearance, 194, 240, 251, 264, 273
Mucociliary functions, 198, 204, 205, 209, 237
Mucostasis, 195, 198, 201, 205, 206, 237
Mucus:
  composition of, 203, 251
  flow of, 195, 197-199, 201 204, 235, 269
  general, 194
  secretion of, 263
  secretions and carcinogenesis, 261
Mutagenicity, formaldehyde, 264
Mutation frequencies, 178

N

Nasal irritation by acetaldehyde, 274
Nasal mucosa and carcinogenicity, 261
Nasal turbinates:
    levels, 228
    squamous cell carcinomas of, 32, 168, 194, 226, 240, 242
Nodules, skin, 165, 166
No observed effect level, 34, 204, 242
Nucleophiles, 211

O

Odds ratio:
  adjusted, 59, 63, 67, 71
  crude, 74, 75, 78-88, 90, 92-96, 109, 113

relative, 58, 59, 68, 72, 74, 75, 97-101

P

Papillomas, skin, 147, 149-150, 153-155
Phorbol acetate, 147-150, 152-154, 157, 159, 161, 163, 166, 168, 169
Promoting agents (*see* Promotion, cancer)
Promotion, cancer, 147, 148, 151-153, 155-159, 163, 166, 169, 191, 225, 243, 263, 266

R

Risk evaluation:
quantitive, 237-239
regulatory agency practices, 34, 40, 42, 44, 238
Risk ratio, relative, 59

S

Salmonella Typhimurium TM 667, 178, 179, 189
Sister chromatid exchange, 174
Skin painting, 264
Smoking histories, 57, 63, 92, 103, 107
Standardized mortality ratio, 131, 132, 134, 135

T

10T1/2 Mouse fibroblast, 266
Threshold for carcinogenicity, 33, 35, 36, 38, 39, 45
Thymidine kinase, 175, 176
Time-to-tumor, 44, 45
TK locus, 187
Trifluorothymidine resistance, 174, 177, 180

U

UF-insulated home, 38

V

Vinyl chloride, 239, 247
Virtually safe dose, 243

W

Work histories, 24, 55, 70